REVOLVING VECTORS

with Special Application to
Alternating Current Phenomena

by

George W. Patterson, Ph.D.

Professor of Electrical Engineering
University of Michigan

Electrical Engineering Series

Wexford Press
2008

PREFATORY NOTE

THE use of complex quantities, i.e., quantities part real and part imaginary, in the theory of alternating currents has been greatly developed by Dr. Charles P. Steinmetz in his work on "Alternating Current Phenomena." It would be difficult to determine the influence which earlier writers, from the time of Caspar Wessel down to Steinmetz's day, have had in laying the foundation on which Steinmetz has built. It is, however, fair to say that the great advance in the use of vector methods, both algebraic and geometric, due to Dr. Steinmetz, justifies us in calling their application to alternating current phenomena Steinmetz's Method.

Earlier writers used complex quantities to represent *vector* quantities algebraically. Dr. Steinmetz extended the application so as to include *harmonic* quantities. As many writers on electrical subjects are prone to confuse vector and harmonic quantities, the author thinks it necessary to distinguish these two uses of complex quantities, and for that purpose he starts with the *vector* use and later takes up the *harmonic* use. In addition, subtraction and certain cases of multiplication and division, correct results are obtained by treating harmonic quantities as vector quantities; but in other cases of multiplication (such as multiplication of e.m.f. and current to obtain power) and division (such as dividing power by e.m.f. to get current) incorrect results are obtained unless arbitrary rules of multiplication and division are introduced. It therefore is necessary thoroughly to examine the fundamentals of these uses of complex quantities, and to deduce the laws of addition, subtraction, multiplication, and division, as applicable to vector quantities and to harmonic quantities whether simple (electromotive force and current) or compound (power), and also to such non-harmonic quantities as resistance, capacity, inductance, etc., in connection with harmonic quantities.

There are two methods employed by electricians using the complex quantity notation. The older method due to Dr. Steinmetz is expressed in graphical form by the wave diagram. The other method uses the so-called crank diagram. In both methods counter-clockwise rotations are used, though the formulæ have led some persons to think that Dr. Steinmetz has used clockwise rotations. It is true that the imaginary terms in the resulting formulæ have opposite signs. In reading Dr. Steinmetz's works no confusion need result for one accustomed to the crank diagram method if the differences are kept in mind. It has seemed to the author that the crank diagram method suits his purpose better, and consequently it will be used in this book.

TABLE OF CONTENTS

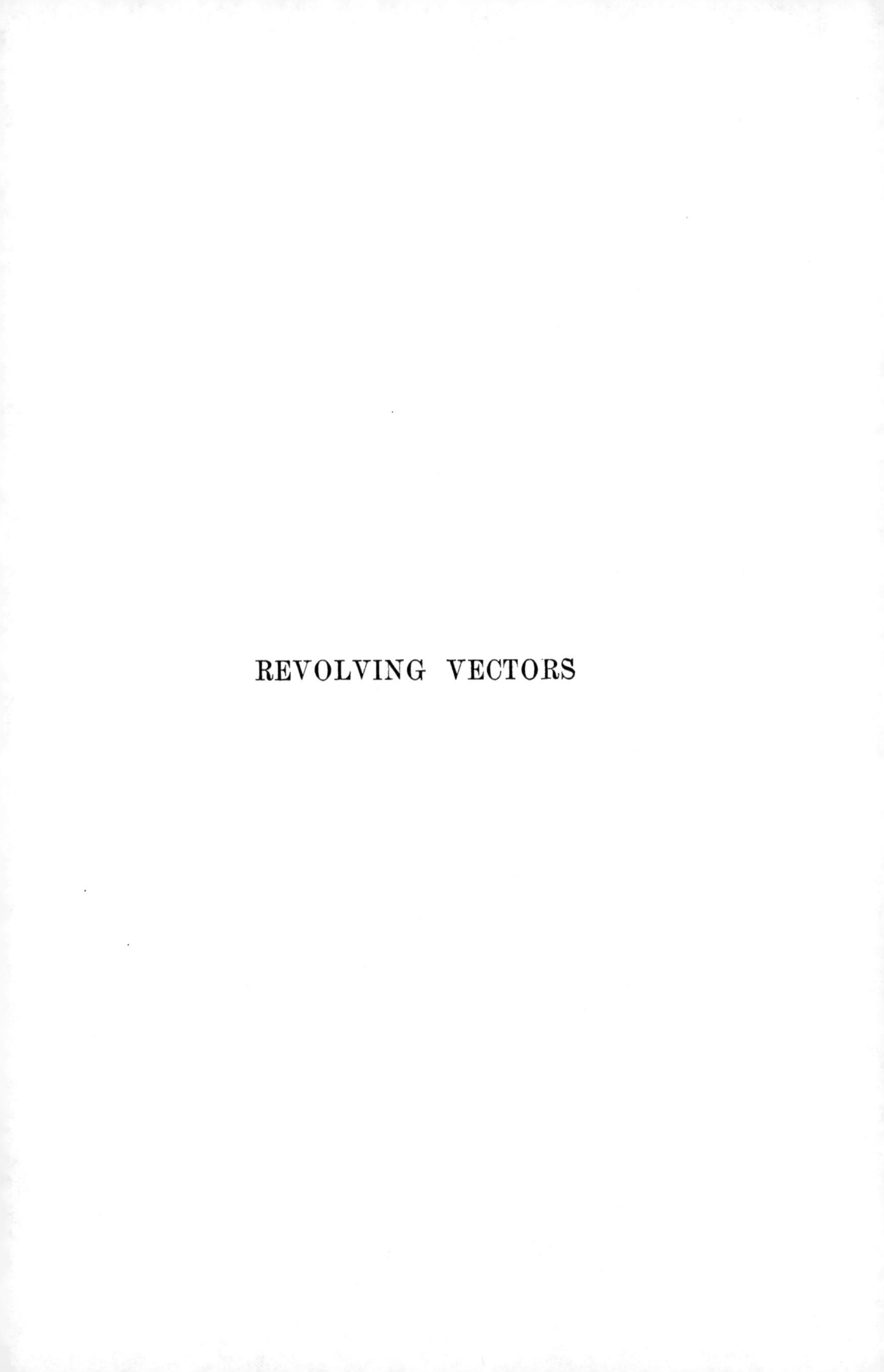

REVOLVING VECTORS

REVOLVING VECTORS

CHAPTER I

ROTARY POWER OF ROOTS OF MINUS ONE

§ 1. Before the year 1797 algebraic expressions were used to represent magnitudes only, but in that year a Danish surveyor, Caspar Wessel, presented a memoir to the Royal Academy of Sciences and Letters of Denmark, entitled "On the Analytical Representation of Direction[1]." Wessel's memoir laid the foundation for vector anaylsis, for the theory of functions of a complex variable, and for Steinmetz's method for alternating current phenomena. In this paper Wessel introduced $\sqrt{-1}$ as *the sign of perpendicularity*, and used the letter ε to indicate this use of the imaginary unit. It is now common to use the letters i or j for $\sqrt{-1}$. He showed how a quantity might be represented both in magnitude and direction by an algebraic expression, and made it clear that he used *the sign of perpendicularity* not as a factor in the strict algebraic sense, but rather as an operator functioning to rotate the mag-

[1] Om Direktionens analytiske Betegning, et Forsög anvendt fornemmelig til plane og sphaeriske Polygoners Oplösning; read March 10, 1797; Memoirs of the Academy, Vol. V, 1799; republished in French by the Academy 1897; see also Beman, Proc. A. A. A. S., 1897, Vol. XLVI, p. 33.

nitude for which the rest of the algebraic quantity stood, through an angle of 90°.

Until the discovery by Wessel of the use of $\sqrt{-1}$ as the sign of perpendicularity, this symbol when occurring in a problem had always been taken as a sign indicating the impossibility of the problem, just as we are accustomed to view the answer obtained as absurd if, for example, we are required to divide seven apples into piles having ten apples in each. In early ages minus quantities were unknown, and solutions involving the indication of a minus quantity were taken to indicate an impossibility. We now think it fair to hold our minds open to some new meaning to be assigned to symbols which formerly were meaningless if not absurd.

Wessel's memoir met the same fate as many others written in advance of their time, such as Green's essay in which the potential function was christened, and Gibbs's essays in which the foundation of the thermodynamics of the voltaic cell was laid; for Wessel's paper was put to sleep in the printed memoirs of the Academy, its slumber not to be disturbed until long years after, when Wessel's ideas had been rediscovered by other men such as Argand, Gauss, Cauchy, Français, and Gergonne.

§ 2. To show how little prepared the mathematical world was for Wessel's use of $\sqrt{-1}$, it is interesting to find that Cauchy [1] as late as 1844 said:

> "Every imaginary equation is naught else than the symbolic representation of *two* equations between real quantities. The employment of imaginary expressions by permitting us to replace two equations by a single one, often offers the means of simplifying calculations and of writing in abridged form quite complicated results. Such indeed is the principal motive for con-

[1] Cauchy, Exercise d'analyse et de physique mathematique, Tome III, p. 361.

tinuing our use of these expressions, *which taken literally and interpreted according to generally established conventions signify nothing and are without sense.*"

Professor Durège of Prague says in the introduction to his book on "The Theory of Functions of a Complex Variable"[1]:

> "The work of deMoivre, Bernoulli, the two Fagnanos, d'Alembert, Euler, and others was, on the whole, looked upon more as scientific foolery (Spielereien für blose Curiosa), and that it was entitled to appreciation of worth only in proportion as it lent useful means to help in other investigations."

§ 3. To prepare for a meaning to an even root of a negative quantity, it may be useful to consider how a negative quantity was transferred from the absurd and impossible to the category of real and possible quantities. We are all in agreement that no quantity, in the strict sense, can be a quantity at all if its magnitude is less than zero. How are we then to understand the negative sign, if a negative quantity is to have real meaning? The solution of the puzzle is illustrated by means of such a problem as this: The point A is five miles east of B, the point C is ten miles east of B. How many miles is A east of C? The result is -5 miles. The old interpretation of the result is that the answer is absurd, for A is not east of C at all. The modern intepretation is that the answer is not absurd and that the -5 miles is to be understood as 5 miles west. Or put in another way, the negative sign is not to be understood as compelling us to consider a distance less than nothing, but simply that the minus sign is an operator which functions to change our eastward sense of counting into the opposite or westward.

[1] Durège, "Theorie der Funktionen einer complexen veränderlichen Grösse," p. 2.

We may then consider that what has been taken as multiplication by -1 is not really multiplication at all, but merely the use of an operator turning an eastward through 180° into a westward sense of counting.

§ 4. In an analogous way, let us consider what would happen if an operator could be found which on being applied to eastward sense would change it to northward and on being applied a second time would change northward to westward, and so on with successive applications of this operator, changing westward to southward, and southward to eastward. That $\sqrt{-1}$ is such an operator was discovered by Wessel, and as before mentioned, he called it *the sign of perpendicularity*, for he found that $\sqrt{-1}$ used twice would cause a reversal of direction or a rotation of 180°; and what was more natural than to assume that one application would produce a rotation of 90°? To avoid ambiguity between the two senses in a plane through which the 180° rotation to produce a reversal might be taken, we may agree to consider the rotation to take place counter-clockwise. We thus have

$$\sqrt{-1}\ \text{East} = \text{North},$$

$$\sqrt{-1}\sqrt{-1}\ \text{East} = \sqrt{-1}\ \text{North} = \text{West},$$

$$\sqrt{-1}\sqrt{-1}\sqrt{-1}\ \text{East} = \sqrt{-1}\sqrt{-1}\ \text{North}$$
$$= \sqrt{-1}\ \text{West} = \text{South},$$

$$\sqrt{-1}\sqrt{-1}\sqrt{-1}\sqrt{-1}\ \text{East} = \sqrt{-1}\sqrt{-1}\sqrt{-1}\ \text{North}$$
$$= \sqrt{-1}\sqrt{-1}\ \text{West} = \sqrt{-1}\ \text{South} = \text{East}.$$

To abridge the notation, but not compromising our attitude, we may write;

$$(\sqrt{-1})^4\ \text{East} = (\sqrt{-1})^3\ \text{North}$$
$$= (\sqrt{-1})^2\ \text{West} = \sqrt{-1}\ \text{South} = \text{East}.$$

§ 5. Although no contradictory complication has resulted from the use of $\sqrt{-1}$ as a 90° rotator, the reader may have grave doubts of the safety of using it in all cases as Wessel's sign of perpendicularity. His confidence may be increased by showing that analogous assumptions of rotary powers, given to other roots of -1, are free from contradictions. To this end let us examine $\sqrt[3]{-1}$, which we should expect to be endowed with the ability to rotate through 60°, as three applications, *as an algebraic multiplier*, would be equivalent to multiplying by -1. How about it as an *operator?* Let us assume that

$$x=\sqrt[3]{-1},$$

or

$$x^3+1=0.$$

This equation has three roots, as follows:

$$x=\tfrac{1}{2}+\tfrac{1}{2}\sqrt{3}\sqrt{-1},$$

$$x=-1,$$

$$x=\tfrac{1}{2}-\tfrac{1}{2}\sqrt{3}\sqrt{-1}.$$

Assuming that $\sqrt[3]{-1}$ may be used as an operator to an eastward direction as before, we have three results:

$$\sqrt[3]{-1}\text{ East}=(\tfrac{1}{2}+\tfrac{1}{2}\sqrt{3}\sqrt{-1})\text{ East}=\tfrac{1}{2}\text{ East}+\tfrac{1}{2}\sqrt{3}\text{ North},$$

or

$$\sqrt[3]{-1}\text{ East}=-1\text{ East}=\text{West},$$

or

$$\sqrt[3]{-1}\text{ East}=(\tfrac{1}{2}-\tfrac{1}{2}\sqrt{3}\sqrt{-1})\text{ East}=\tfrac{1}{2}\text{ East}+\tfrac{1}{2}\sqrt{3}\text{ South}.$$

From Fig. 1 it is evident that the first result produces a rotation of 60° without change in magnitude, for the sine of 60° is $\tfrac{1}{2}\sqrt{3}$ and the cosine is $\tfrac{1}{2}$. The second result is a rotation of 180°, simply changing eastward into westward. The third result is either a backward rotation of

60° or a forward (counter-clockwise) rotation of 300°. All on being repeated for the third time produce reversals of direction; for they give one-half a rotation, one and one-half rotations, and two and one-half rotations respectively, all being taken as counter-clockwise.

To avoid confusion we shall take $\sqrt[3]{-1}$ to be an operator endowed with the power of producing a rotation of 60°.

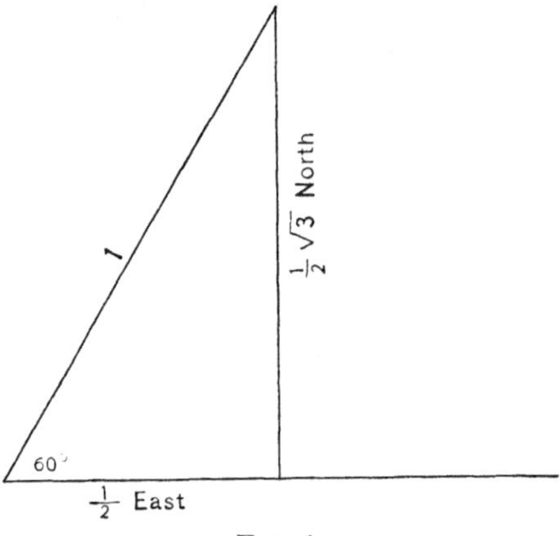

Fig. 1.

§ 6. We may proceed in a similar way to show that $\sqrt[4]{-1}$ may be used to produce a rotation of 45°. Let

$$x=\sqrt[4]{-1},$$

or

$$x^4+1=0.$$

This equation has four solutions, as follows:

$$x=\ \ \tfrac{1}{2}\sqrt{2}+\tfrac{1}{2}\sqrt{2}\sqrt{-1},$$

$$x=-\tfrac{1}{2}\sqrt{2}+\tfrac{1}{2}\sqrt{2}\sqrt{-1},$$

$$x=-\tfrac{1}{2}\sqrt{2}-\tfrac{1}{2}\sqrt{2}\sqrt{-1},$$

$$x=\ \ \tfrac{1}{2}\sqrt{2}-\tfrac{1}{2}\sqrt{2}\sqrt{-1}.$$

Applying them successively to an eastward direction as before we have the four results,

$$\sqrt[4]{-1}\ \text{East} = \tfrac{1}{2}\sqrt{2}\ \text{East} + \tfrac{1}{2}\sqrt{2}\ \text{North} = \text{Northeast},$$

or $$= \tfrac{1}{2}\sqrt{2}\ \text{West} + \tfrac{1}{2}\sqrt{2}\ \text{North} = \text{Northwest},$$

or $$= \tfrac{1}{2}\sqrt{2}\ \text{West} + \tfrac{1}{2}\sqrt{2}\ \text{South} = \text{Southwest},$$

or $$= \tfrac{1}{2}\sqrt{2}\ \text{East} + \tfrac{1}{2}\sqrt{2}\ \text{South} = \text{Southeast}.$$

They are equivalent to rotations of 45°, 135°, 225° or 315°, and all on being repeated for the fourth time produce reversals (without or with extra complete rotations).

§ 7. In the same way all other roots of -1 may be examined (for all are known), and in every case these roots will be found to be endowed with the property that as operators they will produce rotations which, on being repeated to the number of times indicated by the index of the root, will produce reversals. In general for the root giving smallest rotation

$$\sqrt[n]{-1} = \cos\frac{\pi}{n} + \sqrt{-1}\sin\frac{\pi}{n}.$$

Indicating $\sqrt{-1}$ by j, we may write this equation

$$j^{\frac{2}{n}} = \cos\frac{\pi}{n} + j\sin\frac{\pi}{n}.$$

There are also other roots giving larger rotations. The general expression for all roots is as follows:

$$j^{\frac{2}{n}} = \cos\frac{(2m+1)\pi}{n} + j\sin\frac{(2m+1)\pi}{n},$$

where m stands for zero or any positive or negative whole number. It will be shown below that n need not be a positive whole number. In fact it may be positive or negative, a whole number, or a proper or improper fraction, or indeed any number whatever, and we shall even make use of it as a variable quantity in alternating current applications.

§ 8. Let us consider two operators, able to produce rotation through angles indicated by θ and ϕ,

$$A=\cos\theta+j\sin\theta,$$

$$B=\cos\phi+j\sin\phi.$$

Multiplying them together, we get an operator $A\cdot B$,

$$A.B=\cos\theta\cos\phi-\sin\theta\sin\phi+j(\sin\theta\cos\phi+\cos\theta\sin\phi)$$

$$=\cos(\theta+\phi)+j\sin(\theta+\phi).$$

This new operator has the power of producing a rotation through the sum of the angles $\theta+\phi$. It is to be remembered that $j^2=-1$.

If we divide one by the other, we obtain

$$\frac{A}{B}=\frac{\cos\theta+j\sin\theta}{\cos\phi+j\sin\phi}$$

$$=\frac{\cos\theta\cos\phi+\sin\theta\sin\phi+j(\sin\theta\cos\phi-\cos\theta\sin\phi)}{\cos^2\phi+\sin^2\phi}$$

$$=\cos(\theta-\phi)+j\sin(\theta-\phi).$$

From which it appears that $\frac{A}{B}$ is an operator producing a rotation through the angle $\theta-\phi$.

In a similar way to the multiplication above, if θ and ϕ are equal, we have

$$A^2=\cos 2\theta+j\sin 2\theta,$$

$$A^3=\cos 3\theta+j\sin 3\theta\cdot$$

or in general,

$$A^n=\cos n\theta+j\sin n\theta=(\cos\theta+j\sin\theta)^n.$$

If $n\theta =$ equals 180° or π, we have

$$A^n = A^{\frac{\pi}{\theta}} = \cos \pi + j \sin \pi = -1,$$

and

$$A = \sqrt[n]{-1}.$$

From this follows the value of $\sqrt[n]{-1}$ given above, viz.,

$$\sqrt[n]{-1} = \cos \frac{\pi}{n} + j \sin \frac{\pi}{n}.$$

The general value given above may be reached in an analogous manner, for if m is any whole number (positive or negative) or zero, we have

$$\cos (2m+1)\pi + j \sin (2m+1)\pi = -1,$$

and

$$\sqrt[n]{-1} = \cos \frac{(2m+1)\pi}{n} + j \sin \frac{(2m+1)\pi}{n}.$$

This last expression, though it appears to have an indefinitely great number of different values, in fact has only n different values, if n is a whole number; for the values of the cosine and sine repeat after n different values, it being evident that if $m = n$,

$$\cos \frac{(2n+1)\pi}{n} + j \sin \frac{(2n+1)\pi}{n} = \cos \frac{\pi}{n} + j \sin \frac{\pi}{n}.$$

If $m = n + a$,

$$\cos \frac{(2n+2a+1)\pi}{n} + j \sin \frac{(2n+2a+1)\pi}{n} = \cos \frac{(2a+1)\pi}{n} + j \sin \frac{(2a+1)\pi}{n},$$

and so on.

§ 9. If n is a fraction equal to $\frac{p}{q}$, which may be a proper or improper fraction and positive or negative, both p and q being whole numbers, we shall have

$$\sqrt[\frac{p}{q}]{-1}=j^{\frac{2q}{p}}=\cos\frac{q\pi}{p}+j\sin\frac{q\pi}{p};$$

for on raising this expression to the power p, we obtain

$$j^{2q}=\cos q\pi+j\sin q\pi.$$

If q is an odd number $j^{2q}=\cos q\pi=-1$, and if q is an even number $j^{2q}=\cos q\pi=+1$. In both cases $\sin q\pi$ is zero. If q is negative the same result follows.

§ 10. If n is a number which is neither whole nor a proper or improper fraction, we may by the doctrine of limits have confidence in assuming that $\sqrt[n]{-1}$ will have a value between $\sqrt[r]{-1}$ and $\sqrt[s]{-1}$, where $r<n<s$, and r and s are whole numbers or fractions very close to one another in value.

As we may always find whole numbers or fractions, one larger and one smaller and different from n by amounts less than any assigned amount, in the limit we may find the value of $\sqrt[r]{-1}$ with as high a degree of precision as desired. We may therefore have confidence that n may be a continuously varying quantity, say a function of the time. For example let ω be an angular velocity and t be time. We may write the following equation:

$$\sqrt[\frac{\pi}{\omega t}]{-1}=j^{\frac{2\omega t}{\pi}}=\cos\omega t+j\sin\omega t.$$

This equation expresses a *variable operator* which functions to rotate any vector to which it is applied with a counter-clockwise angular velocity ω. In the case of clockwise rotation, substituting $-\omega$ for ω. we obtain

$$\sqrt[-\frac{\pi}{\omega t}]{-1}=j^{-\frac{2\omega t}{\pi}}=\cos\omega t-j\sin\omega t.$$

It should be had in mind that reversing the sign of an angle does not affect the cosine, but does reverse the sine.

§ 11. The formulæ for rotating operators, as will be shown in the next chapter, may be more conveniently expressed as powers of the base ε of the Napierian system of logarithms, as follows:

$$\varepsilon^{j\omega t} = \cos \omega t + j \sin \omega t,$$

and

$$\varepsilon^{-j\omega t} = \cos \omega t - j \sin \omega t.$$

CHAPTER II

ROTARY POWER OF IMAGINARY EXPONENTS

§ 12. It has been shown in the previous chapter that powers or roots of the imaginary unit, $\sqrt{-1}$, or j, may be used to obtain an operator which can function to rotate a vector quantity either through a stated angle or through an angle increasing continuously with the time. These operators may be more conveniently expressed as imaginary powers of ε, the base of the Napierian system of logarithms. The mathematical formulæ involved were already old and well known in Wessel's day and are to be found in Euler's memoirs.

If we expand ε^x, $\sin x$ and $\cos x$ in powers of x we obtain:

$$\varepsilon^x = 1 + \frac{x}{\lfloor 1} + \frac{x^2}{\lfloor 2} + \frac{x^3}{\lfloor 3} + \frac{x^4}{\lfloor 4} + \frac{x^5}{\lfloor 5} + \text{etc.}\ \frac{x^n}{\lfloor n} + \text{etc.}$$

$$\sin x = \frac{x}{\lfloor 1} - \frac{x^3}{\lfloor 3} + \frac{x^5}{\lfloor 5} - \frac{x^7}{\lfloor 7} + \text{etc.} + (-1)^n \frac{x^{2n+1}}{\lfloor 2n+1} + \text{etc.}$$

$$\cos x = 1 - \frac{x^2}{\lfloor 2} + \frac{x^4}{\lfloor 4} - \frac{x^6}{\lfloor 6} + \text{etc.} + (-1)^n \frac{x^{2n}}{\lfloor 2n} + \text{etc.}$$

It is evident that if jx is substituted for x we shall have

$$\varepsilon^{jx} = 1 + j\frac{x}{\lfloor 1} - \frac{x^2}{\lfloor 2} - j\frac{x^3}{\lfloor 3} + \frac{x^4}{\lfloor 4} + j\frac{x^5}{\lfloor 5} + \text{etc.}$$

$$= \cos x + j \sin x.$$

Referring to Chapter I it is evident that the rotating operator may take any one of four equal forms:

$$\sqrt[\frac{\pi}{\theta}]{-1}=j^{\frac{2\theta}{\pi}}=\varepsilon^{j\theta}=\cos\theta+j\sin\theta,$$

in case rotation is to take place through a definite angle θ; or in case the rotation is to be continuous and a function of the time the four equal forms may be written:

$$\sqrt[\frac{\pi}{\omega t}]{-1}=j^{\frac{2\omega t}{\pi}}=\varepsilon^{j\omega t}=\cos\omega t+j\sin\omega t,$$

in which t is the time and ω the angular velocity.

§ 13. If an operator consisting of the sum of two operators which used singly would produce rotations equal in magnitude but opposite in sense, is used on a vector, the operator reduces to a simple factor causing the vector to follow the law of simple harmonic motion. The expression for such an operator is:

$$\varepsilon^{j\omega t}+\varepsilon^{-j\omega t}=2\cos\omega t,$$

a result which might have been obtained directly from Euler's formula for the cosine,

$$\cos\theta=\frac{\varepsilon^{j\theta}+\varepsilon^{-j\theta}}{2}.$$

CHAPTER III

POSITION OF A POINT IN A PLANE

§ 14. Many applications of complex quantities with a vector meaning might be made. It is believed, however, that the following will suffice to illustrate their use.

The position of a point in a plane may be determined by rectangular coordinates. Using as before j for Wessel's sign of perpendicularity, the position of the point P with reference to the point O taken as the origin of coordinates as shown in Fig. 2, has the following expression:

$$P = a + jb,$$

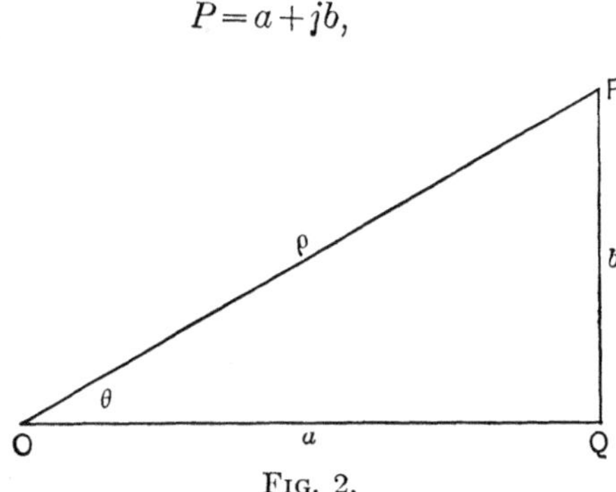

FIG. 2.

a being the horizontal and b the vertical projection of the line of length ρ, connecting O and P. We have by geometry that $\rho^2 = a^2 + b^2$, $a = \rho \cos \theta$, and $b = \rho \sin \theta$. Using polar coordinates, the equation becomes

$$\boldsymbol{P} = \boldsymbol{\rho} \cos \theta + j\rho \sin \theta = (\cos \theta + j \sin \theta)\rho.$$

Using the notation of the last chapter, we have

$$P=\varepsilon^{j\theta}\rho=(\cos\theta+j\sin\theta)\rho.$$

Wessel's idea of the analytical representation of direction has an illustration in each of the equal operators $\varepsilon^{j\theta}$ and $\cos\theta+j\sin\theta$. Each has a magnitude unity, and each may be considered to have the sole effect of specifying a direction differing by an angle θ from the direction (horizontal) taken as standard. It is permissible also, as previously done, to consider $\varepsilon^{j\theta}$ and $\cos\theta+j\sin\theta$ as operators functioning to turn ρ from a horizontal position to that indicated in the figure. The whole expression for P,

$$P=a+jb=\varepsilon^{j\theta}\rho=(\cos\theta+j\sin\theta)\rho,$$

has both magnitude and direction expressed, the first form, $a+jb$, expressing by a both magnitude and horizontal direction and by b both magnitude and horizontal direction, the latter being rotated into a vertical direction by the operation of j. In the later expressions the operators, or analytical expressions of direction, $\varepsilon^{j\theta}$ and $\cos\theta+j\sin\theta$, are expressed separately from the magnitude ρ which, if standing alone, would have been understood as a horizontal magnitude.

UNIFORM CIRCULAR MOTION

§ 15. Another simple illustration may be taken from uniform circular motion, one of the simplest motions met with in physics and which we shall use later in connection with harmonic quantities.

First let us consider the position of a point moving about the circumference of a circle. Let the radius of the circle be R and the angular position of the radius be expressed in terms of angular velocity ω and the time t elapsed since the radius was horizontal and directed to the right (Fig. 3). Let us take the center of the circle O as origin of coordinates. We have then

$$P=\varepsilon^{j\omega t}R=(\cos\omega t+j\sin\omega t)R.$$

If for any reason it is desirable to take as origin any other point with coordinates a and jb, and to measure angular position from a radius making an angle θ with the horizontal, we would have

$$P=\varepsilon^{j(\omega t-\theta)}R-a-jb=(\cos(\omega t-\theta)+j\sin(\omega t-\theta))R-a-jb.$$

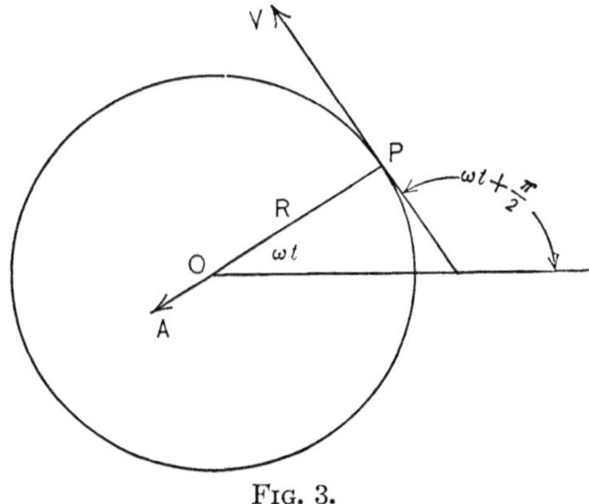

Fig. 3.

The introduction of an eccentric origin (a, jb) and an epoch (θ) introduces no real difficulty, though it complicates the expression.

Consider now the velocity V of the point P. Evidently we have (using the earlier expression for P),

$$V=\frac{dP}{dt}=j\omega\varepsilon^{j\omega t}R=\omega\left(\cos\left(\omega t+\frac{\pi}{2}\right)+j\sin\left(\omega t+\frac{\pi}{2}\right)\right)R$$

$$=\varepsilon^{j\left(\omega t+\frac{\pi}{2}\right)}\omega R=\left(\cos\left(\omega t+\frac{\pi}{2}\right)+j\sin\left(\omega t+\frac{\pi}{2}\right)\right)\omega R.$$

This shows that the magnitude of the velocity is ωR, and the phase 90° or $\frac{\pi}{2}$ ahead of the phase of P, both well-known facts of uniform circular motion. As the last transformation may have difficulties for some readers, it is well to note that, as given earlier,

$$j^{\frac{2\omega t}{\pi}} = \varepsilon^{j\omega t},$$

and if ωt equals $\frac{\pi}{2}$, this equation reduces to

$$j = \varepsilon^{j\frac{\pi}{2}},$$

therefore

$$j\varepsilon^{j\omega t} = \varepsilon^{j\left(\omega t + \frac{\pi}{2}\right)}.$$

It is also of advantage in differentiating $\cos \omega t + j \sin \omega t$ not to change from cosine to sine and vice versa, but rather to advance the phase by $\frac{\pi}{2}$, which comes to the same thing. Changing from cosine to sine and vice versa in differentiating harmonic quantities conceals the change of phase from immediate notice, and a clear understanding of phase relations is desirable.

If the origin of coordinates is not at the center of the circle and if there is an epoch angle in the expression, the second expression for P,

$$P = \varepsilon^{j(\omega t - \theta)} R - a - jb = (\cos(\omega t - \theta) + j \sin(\omega t - \theta)) R - a - jb$$

leads to a value of V,

$$V = \frac{dP}{dt} = \varepsilon^{j\left(\omega t + \frac{\pi}{2} - \theta\right)} \omega R$$

$$= \left(\cos\left(\omega t + \frac{\pi}{2} - \theta\right) + j \sin\left(\omega t + \frac{\pi}{2} - \theta\right)\right) \omega R.$$

The acceleration A in uniform circular motion is as follows:

$$A=\frac{dV}{dt}=\varepsilon^{j(\omega t+\pi)}\omega^2R=(\cos(\omega t+\pi)+j\sin(\omega t+\pi))\omega^2R,$$

for the simple case, and

$$A=\varepsilon^{j(\omega t+\pi-\theta)}\omega^2R=(\cos(\omega t+\pi-\theta)+j\sin\omega t+\pi-\theta))\omega^2R,$$

for the more involved case. The meaning of the formulæ is that the magnitude of the acceleration is ω^2R; and its phase $\omega t+\pi$ (or $\omega t+\pi-\theta$) shows, by the added π, that the acceleration is directed toward the center.

EFFECT OF DAMPING, SPIRAL MOTION

§ 16. Another interesting example is found in the exponential spiral which a pendulum started in motion in a horizontal circular path will follow if its motion is damped in proportion to its velocity (Fig. 4). The equation for the position of the pendulum is as follows, if we take the center as origin and assume the epoch as zero:

$$P=\varepsilon^{(j\omega-\alpha)t}R=(\cos\omega t+j\sin\omega t)R\varepsilon^{-\alpha t}.$$

In this expression $R\varepsilon^{-\alpha t}$ is the magnitude of the distance from the origin, and $\varepsilon^{j\omega t}$, or $\cos\omega t+j\sin\omega t$, is the analytical expression for the direction of the line from O to P. Differentiating P with respect to t, we get the velocity of the pendulum,

$$V=\frac{dP}{dt}=\left[\left(\cos\left(\omega t+\frac{\pi}{2}\right)+j\sin\left(\omega t+\frac{\pi}{2}\right)\right)\omega - (\cos\omega t+j\sin\omega t)\alpha\right]R\varepsilon^{-\alpha t}=(j\omega-\alpha)\varepsilon^{j\omega t-\alpha t}R.$$

Let $\tan\phi=\frac{\alpha}{\omega}$, $\sin\phi=\frac{\alpha}{\sqrt{\alpha^2+\omega^2}}$ and $\cos\phi=\frac{\omega}{\sqrt{\alpha^2+\omega^2}}$.

The expression for V then may be rewritten

$$V=\left[\cos\left(\omega t+\frac{\pi}{2}+\phi\right)+j\sin\left(\omega t+\frac{\pi}{2}+\phi\right)\right]R\sqrt{\alpha^2+\omega^2}\varepsilon^{-\alpha t}$$

$$=\varepsilon^{j\left(\omega t+\frac{\pi}{2}+\phi\right)-\alpha t}R\sqrt{\alpha^2+\omega^2}.$$

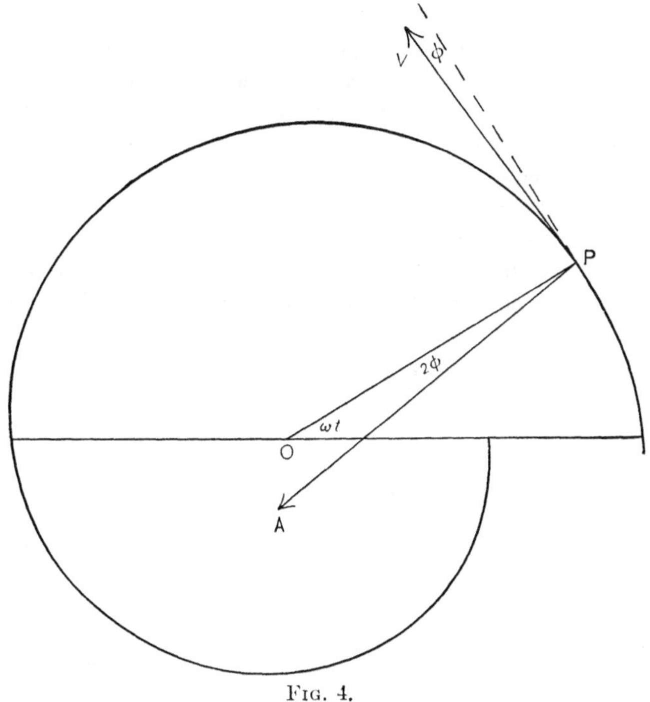

FIG. 4.

The acceleration A becomes

$$A=[\cos\omega t+\pi+2\phi)+j\sin(\omega t+\pi+2\phi)]R(\alpha^2+\omega^2)\varepsilon^{-\alpha t}$$

$$=\varepsilon^{j(\omega t+\pi+2\phi)-\alpha t}R(\alpha^2+\omega^2).$$

These equations show that the phase of the velocity of a damped circular pendular motion is $\frac{\pi}{2}+\phi$ in advance of the phase of the position of the pendulum, *i.e.*, ϕ more than a quadrant: and the phase of the acceleration is in advance of that of the velocity an equal amount. The acceleration is not directed toward the center, as is the case in uniform circular motion, but is in advance of a line drawn to the center by an angle 2ϕ.

The real part of the above expressions is applicable to simple pendular motion (in a vertical plane) or to the movement of a ballistic galvanometer with damping of moderate magnitude, and analogous expressions apply to the charge and current in the case of the oscillatory discharge of a condenser in an inductive circuit.

The origin need not be taken at the center of the spiral, and there may be an epoch angle if for any reason it is considered desirable not to take the coordinates as assumed above. The complications resulting are not troublesome.

CHAPTER IV

SIMPLE HARMONIC QUANTITIES

§ 17. In the previous chapters we have used complex quantities in connection with real vectors only. In this chapter we shall make use of vector expressions to represent simple harmonic quantities.

In Chapter II the connection was shown between a simple harmonic motion and a pair of circular motions equal in magnitude but with oppositely directed angular velocities. Algebraically this connection is expressed by Euler's formula for the cosine,

$$\cos \omega t = \frac{\varepsilon^{j\omega t} + \varepsilon^{-j\omega t}}{2}.$$

As is well known, the real parts of $\varepsilon^{j\omega t}$ and $\varepsilon^{-j\omega t}$ are identical and the imaginary parts equal and opposite. We therefore have the relation

$$\cos \omega t = \frac{\varepsilon^{j\omega t} + \varepsilon^{-j\omega t}}{2} = \text{real part } [\varepsilon^{j\omega t}] = \text{real part } [\varepsilon^{-j\omega t}].$$

It appears from this expression that instead of expressing the simple harmonic motion as the sum of two oppositely directed uniform circular motions which are equal in magnitude, we might equally well have considered the simple harmonic motion as the real part of a single uniform circular motion of twice the magnitude of one of the pair and revolving either clockwise or counter-clockwise as one may prefer. This statement amounts to saying that a simple harmonic motion is the projection of a uniform circular motion on the diameter of the circle, or as many writers say: simple harmonic motion is the apparent motion of a point in uniform circular motion when viewed from a distant point in the plane of the motion.

HARMONIC ELECTROMOTIVE FORCE

§ 18. Let us consider an electromotive force of the form

$$e = E \cos \omega t,$$

which may be considered as the real part of the expression,

$$\underset{\cdot}{E} = \varepsilon^{j\omega t} E = (\cos \omega t + j \sin \omega t) E.$$

This equation is represented graphically by OP, the radius of the circle understood to be revolving counter-clockwise in the figure (Fig. 5).

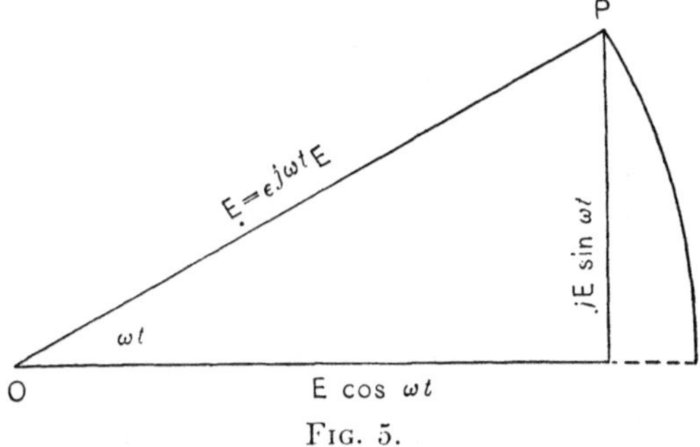

FIG. 5.

A dot over or under a symbol will be understood to mean that the quantity is analogous to a uniform circular motion, but no information is given with respect to the period or phase of the variable. It must always be had in mind that only the real part of the complex expression is to be considered seriously. The other part is to be looked upon as scaffolding about a building in process of erection, or the sawdust in a box of torpedoes, which need not be confused with the building or the torpedoes themselves, respectively, for the imaginary symbol is a warning that the associated term is to be disregarded. Terms in which j occurs as an index must be resolved into their real and imaginary parts before the latter may be disregarded.

No error will be made in adding or subtracting such expressions, for the real part of the sum or the difference of two complex quantities is the sum or the difference of the real parts only. Multiplication or division by, or differentiation with respect to, any real quantity cannot cause any confusion; for none of these processes can change a term from real to imaginary or *vice versa.* But multiplication, or division by, or differentiation with respect to, any imaginary or complex quantity is apt to result in confusion unless quite arbitrary rules are used for these operations. As a rule in the multiplication of two simple harmonic quantities, we may not use the whole expression, but only the real parts. As an example, in obtaining the expression for power by multiplying current and e.m.f., we must use the real parts only. Power, as a rule, is not a simple harmonic quantity, but is a sum (or difference) of a constant and a simple harmonic quantity of double frequency. Dr. Steinmetz by using an arbitrary rule for such multiplication obtains the *average* value of the power. As it may be shown that Dr. Steinmetz's rule for obtaining average power always leads to the right result, his rule may be used fearlessly in such cases.

§ 19. In representing harmonic current or e.m.f. by means of the analytical expression for a revolving vector, it was assumed above that the projection, represented by the real part of the expression, should be the graphical representation of the current or e.m.f. respectively, and therefore diagrams should be drawn to the proper scale. For many purposes it will be found more convenient to change the scale in such a way that the length of the revolving vector shall represent the effective (square root of mean square) value which is indicated by an ammeter or voltmeter in the respective cases. This value for simple harmonic cases is $\frac{1}{2}\sqrt{2}$, about 0.707, times the maximum value. The analogous equation is as follows:

$$\dot{E} = \varepsilon^{j\omega t}\sqrt{2}E = (\cos \omega t + j \sin \omega t)\sqrt{2}E,$$

if the value at any time t is to be given by the projection. As it is only rarely that we desire to know instantaneous values, it is more usual to use the former expression

$$\dot{E} = \varepsilon^{j\omega t}E = (\cos \omega t + j \sin \omega t)E,$$

and understand by E the reading of the voltmeter, and in case *instantaneous* values are ever needed, to find them by multiplying the real part of $\dot{E}$ at any instant by $\sqrt{2}$. The beginner must early master the difficulties introduced or avoided by using or suppressing $\sqrt{2}$ in the formulæ, and be on his guard to avoid misunderstanding various writers. It may be said that, as a rule, harmonic currents or electromotive forces are expressed in *effective* values, while harmonic magnetic fields are expressed in *maximum* values. Power is as a rule expressed in *average* values. The reasons for these apparently arbitrary choices the more advanced student has probably already learned. We cannot take space here to go into the matter further, and must be content with the bare statement.

HARMONIC CURRENT, IMPEDANCE

§ 20. If the current, as well as the electromotive force, follows an harmonic law, and lags behind the e.m.f. by a phase difference represented by the angle θ, we may write

$$i = I \cos (\omega t - \theta),$$

where I is the maximum value of i. If the circuit has a resistance R and an inductance L, and is not complicated by capacity or mutual inductance, and includes no motors or sources of e.m.f., Ohm's law modified for varying currents gives

$$e=Ri+L\frac{di}{dt}=E\cos\omega t.$$

Substituting the value for i, as given above, in the last equation, we obtain

$$E\cos\omega t=RI\cos(\omega t-\theta)+L\omega I\cos\left(\omega t-\theta+\frac{\pi}{2}\right).$$

If this equation is true for any and all times, it is evident that

$$E\sin\omega t=RI\sin(\omega t-\theta)+L\omega I\sin\left(\omega t-\theta+\frac{\pi}{2}\right).$$

From this it follows that the next equation is true,

$$\underset{\bullet}{E}=(\cos\omega t+j\sin\omega t)E=RI(\cos(\omega t-\theta)+j\sin(\omega t-\theta))$$
$$+L\omega I\left(\cos\left(\omega t-\theta+\frac{\pi}{2}\right)+j\sin\left(\omega t-\theta+\frac{\pi}{2}\right)\right),$$

and remembering that $j^2=-1$, we have by simple transformations

$$\underset{\bullet}{E}=(R+jL\omega)(\cos(\omega t-\theta)+j\sin(\omega t-\theta))I=\varepsilon^{j\omega t}E,$$

or

$$\underset{\bullet}{E}=(R+jL\omega)\underset{\bullet}{I}=\varepsilon^{j\omega t}E.$$

Substituting the exponential for the cosine and sine expression, we have

$$\underset{\bullet}{E}=\varepsilon^{j(\omega t-\theta)}(R+jL\omega)I=\varepsilon^{j\omega t}E.$$

The diagram (Fig. 6) shows the relations analytically expressed by the equations. The projection of $\underset{\bullet}{E}$ on the horizontal axis (axis of real values) equals the sum of the projections of $R\underset{\cdot}{I}$ and $jL\omega\underset{\cdot}{I}$, as expressed by the earlier equation,

$$e = E \cos \omega t = RI \cos (\omega t - \theta) + L\omega I \cos\left(\omega t - \theta + \frac{\pi}{2}\right).$$

As $L\omega I \cos\left(\omega t - \theta + \frac{\pi}{2}\right)$ is a quarter period in advance of $RI \cos (\omega t - \theta)$, it is evident that the triangle is a right

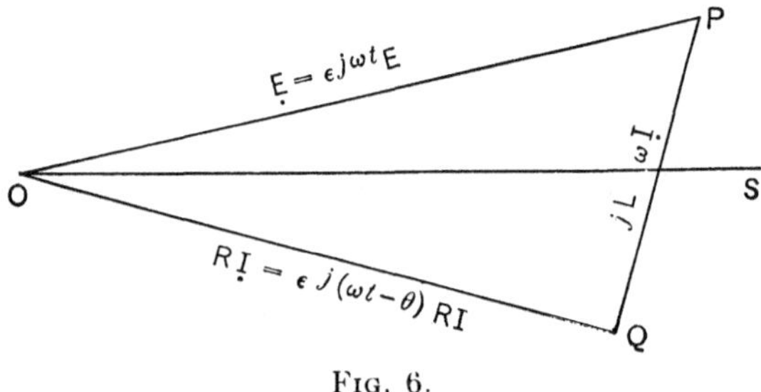

Fig. 6.

triangle. By geometry we then have for the magnitudes involved,

$$E^2 = (R^2 + L^2\omega^2) I^2,$$

and

$$\sqrt{R^2 + L^2\omega^2} = \frac{E}{I}.$$

This ratio between E and I is called the impedance of the circuit and, in the extension of Ohm's law to alternating currents, plays the part that resistance does for direct currents. Impedance is measured in ohms, just as though it were a real resistance.

In a similar way from the uniform circular formula, we have

$$R + jL\omega = \frac{\underset{\cdot}{E}}{\underset{\cdot}{I}} = \frac{\varepsilon^{j\omega t} E}{\varepsilon^{j(\omega t - \theta)} I} = \varepsilon^{j\theta} \frac{E}{I} = \varepsilon^{j\theta} \sqrt{R^2 + L^2\omega^2}.$$

The complex constant $R+jL\omega$ is also called the impedance of the circuit. It evidently has for magnitude $\sqrt{R^2+L^2\omega^2}$, and as an operator rotates an associated quantity counter-clockwise through an angle θ, equal to $\tan^{-1}\dfrac{L\omega}{R}$.

It is perfectly evident that R, ω and L are all *real* quantities and that R and $L\omega$ do not in fact have any quarter-phase relation. To express the *proved* relation between E and I and between $\underset{\cdot}{E}$ and $\underset{\cdot}{I}$, we merely assume the existence of a physical quantity called an impedance, which we express as $\sqrt{R^2+L^2\omega^2}$ or $R+jL\omega$ in the two cases respectively. This is simply a case of the end justifying the means. It is evident, however, that the investigation of the relation between E and I (either effective or maximum values) has been perfectly general, i.e., no special values have been assumed for any of these quantities. We may therefore fearlessly deal with impedances just as though resistances and reactances (as we designate products like $L\omega$) had in fact perpendicular (or quarter-period difference) relations. It must, however, be kept in mind that we are considering simple harmonic quantities and that for other quantities other results follow.

§ 21. Let us now consider the case of a simple circuit with simple harmonic e.m.f. and constant resistance R, inductance L and capacity C all in series. As is well known, after a reasonable time the current reaches its harmonic state and may be expressed as before by the formula

$$i=I\cos(\omega t-\theta).$$

Ohm's law extended to variable conditions gives for the e.m.f.,

$$e=E\cos\omega t=Ri+L\frac{di}{dt}+\frac{1}{C}\int i dt.$$

Substituting the value of i, we have

$$E\cos\omega t=I\left[R\cos(\omega t-\theta)+\left(L\omega-\frac{1}{C\omega}\right)\cos\left(\omega t-\theta+\frac{\pi}{2}\right)\right]+K.$$

It is evident that the constant K must be zero, otherwise the second member of the equation is not simple harmonic. We therefore have

$$E\cos\omega t=I\left[R\cos(\omega t-\theta)+\left(L\omega-\frac{1}{C\omega}\right)\cos\left(\omega t-\theta+\frac{\pi}{2}\right)\right].$$

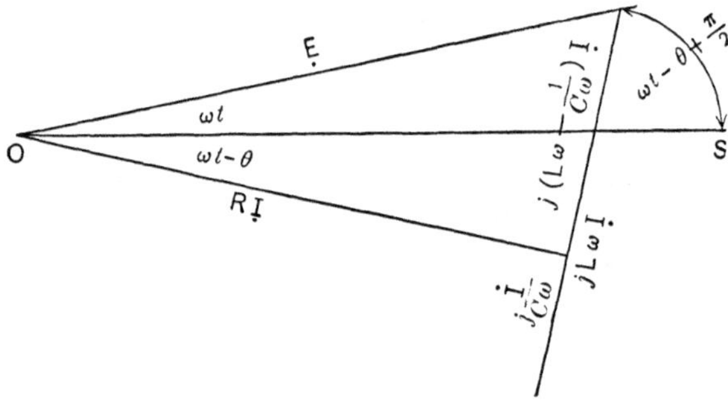

FIG. 7.

This equation is illustrated by Fig. 7. The projections of the sides of the triangle on the horizontal line are evidently

$$E\cos\omega t,\quad IR\cos(\omega t-\theta)\quad\text{and}\quad I\left(L\omega-\frac{1}{C\omega}\right)\cos\left(\omega t-\theta+\frac{\pi}{2}\right).$$

The triangle is understood to be in counter-clockwise rotation about the point O, with an angular velocity ω. From the triangle it is evident that

$$I=\frac{E}{\sqrt{R^2+\left(L\omega-\frac{1}{C\omega}\right)^2}}\quad\text{and}\quad\tan\theta=\frac{L\omega-\frac{1}{C\omega}}{R}.$$

The reactance of the circuit is $L\omega-\frac{1}{C\omega}$. In case $\frac{1}{C\omega}>L\omega$ the angle θ becomes negative, as shown by the diagram (Fig. 8), but the form of the equation remains unchanged. If we consider the projections of the sides of the triangles on a vertical line, we have

$$jE\sin\omega t=jI\left[R\sin(\omega t-\theta)+\left(L\omega-\frac{1}{C\omega}\right)\sin\left(\omega t-\theta+\frac{\pi}{2}\right)\right].$$

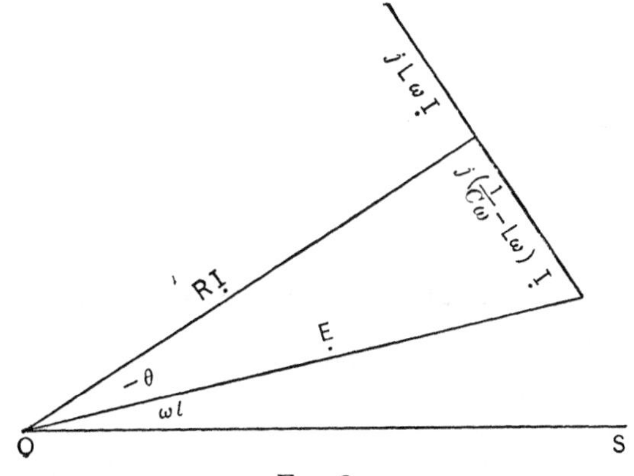

FIG. 8.

Combining these two equations we have, remembering that $\varepsilon^{j\frac{\pi}{2}}=j$,

$$\underset{\cdot}{E}=E\varepsilon^{j\omega t}=RI\varepsilon^{j(\omega t-\theta)}+\left(L\omega-\frac{1}{C\omega}\right)I\varepsilon^{j\left(\omega t-\theta+\frac{\pi}{2}\right)}$$

$$=\left(R+j\left(L\omega-\frac{1}{C\omega}\right)\right)I\varepsilon^{j(\omega t-\theta)}=\underset{\cdot}{I}\left(R+j\left(L\omega-\frac{1}{C\omega}\right)\right)$$

$$=I\sqrt{R^2+\left(L\omega-\frac{1}{C\omega}\right)^2}\,\varepsilon^{j\omega t}.$$

The factor $R+j\left(L\omega-\frac{1}{C\omega}\right)$ and its magnitude $\sqrt{R^2+\left(L\omega-\frac{1}{C\omega}\right)^2}$ are called the impedance of the circuit.

They play the same role as the resistance in the case of unvarying currents for which Dr. Ohm formulated his rule, known as Ohm's law. This factor (in either form) is the ratio between simple harmonic e.m.f. and current, and may be used fearlessly in finding the value of the current with known e.m.f. and *vice versa.* In its complex form impedance indicates not only the magnitude of the ratio, but also the fact that the current lags behind the e.m.f. by an angle θ, whose tangent is expressed by the formula,

$$\tan \theta = \frac{L\omega - \dfrac{1}{C\omega}}{R}.$$

This lag becomes a leading angle if $L\omega < \dfrac{1}{C\omega}$.

HARMONIC ELECTROMOTIVE FORCES IN SERIES

§ 22. The foregoing method may be used for circuits in which two electromotive forces of different phase are in series with a known impedance, or for divided circuits in which the current in different branches has different phases.

As an example of the former, suppose a circuit, of resistance R and reactance $L\omega$, includes two electromotive forces in series, the second lagging behind the first by a phase difference θ. It is assumed that the frequency is the same for both. The combined e.m.f. expressed in a uniform circular formula is

$$\dot{E} = \dot{E}_1 + \dot{E}_2 = \varepsilon^{j\omega t}E_1 + \varepsilon^{j(\omega t - \theta)}E_2 = \varepsilon^{j\omega t}[E_1 + \varepsilon^{-j\theta}E_2]$$

$$= \varepsilon^{j\omega t}[E_1 + (\cos \theta - j \sin \theta)E_2],$$

$$= \varepsilon^{j\omega t}[E_1 + E_2 \cos \theta - jE_2 \sin \theta].$$

Let us assume that $\tan \theta' = \dfrac{E_2 \sin \theta}{E_1 + E_2 \cos \theta}$, and derive corresponding values for sin θ' and cos θ'. We then shall have

$$\dot{E} = \varepsilon^{j\omega t}(\cos\theta' - j\sin\theta')\sqrt{E_1^2+E_2^2+2E_1E_2\cos\theta}$$
$$= \varepsilon^{j(\omega t-\theta')}\sqrt{E_1^2+E_2^2+2E_1E_2\cos\theta},$$

and

$$E = \sqrt{E_1^2+E_2^2+2E_1E_2\cos\theta}.$$

The impedance of the circuit is $R+jL\omega$, and it produces a lag θ'' in the current behind the combined e.m.f. The value of θ'' is $\tan^{-1}\frac{L\omega}{R}$. We have also the relation

$$R+jL\omega = \varepsilon^{j\theta''}\sqrt{R^2+L^2\omega^2}.$$

Therefore the current is

$$\dot{I} = \frac{\dot{E}}{R+jL\omega} = \varepsilon^{j(\omega t-\theta'-\theta'')}\sqrt{\frac{E_1^2+E_2^2+2E_1E_2\cos\theta}{R^2+L^2\omega^2}}.$$

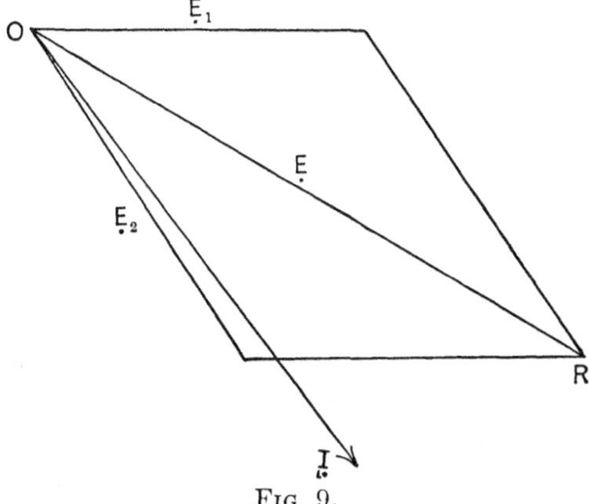

FIG. 9.

Example. Assume $E_1=1000$ volts, $E_2=1200$ volts, $\theta=60°$, $R=12$ ohms, $L\omega=5$ ohms, to find θ', θ'', I, and E.

Answers. $\theta'=\tan^{-1}0.6495$, $\theta''=\tan^{-1}\frac{5}{12}$, $I=146.76$ amp., $E=1907.9$ volts.

The diagram (Fig. 9) indicates graphically the magnitudes and phase relations of E_1, E_2 and E and the phase relation of I. The current is plotted to a different scale, however, to avoid confusion in the diagram. The whole diagram is understood to rotate counter-clockwise about O with an angular velocity $\omega=2\pi f$, when f is the frequency (cycles per second) of the e.m.f.'s.

PROBLEM OF A DIVIDED CIRCUIT

§ 23. The problem of a divided circuit is as follows: Assume an e.m.f. E between junction points of a divided circuit, in one branch of the circuit a resistance R_1, a reactance $L_1\omega$, and a current I_1, in the other branch similar

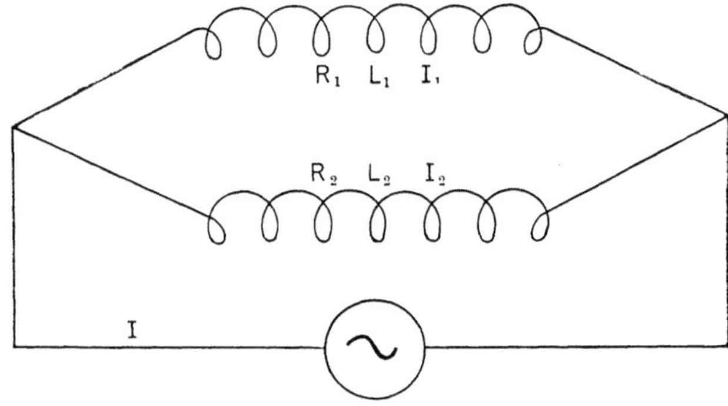

FIG. 10.

quantities R_2, $L_2\omega$ and I_2, and a current I in the undivided part of the circuit. The arrangement of the circuit is as shown in the diagram of connections (Fig. 10). Let us take as known quantities, I_1, R_1, $L_1\omega$, R_2 and $L_2\omega$, and let us find E, I_2 and I, together with the phase relations.

Let us write $\theta_1=\tan^{-1}\dfrac{L_1\omega}{R_1}$ and $\theta_2=\tan^{-1}\dfrac{L_2\omega}{R_2}$.

Let us assume the phase of I_1 to be the standard. We have then

$$\underset{\cdot}{I}_1=(\cos\omega t+j\sin\omega t)I_1=\varepsilon^{j\omega t}I_1,$$

$$\underset{\cdot}{E}=(R_1+jL_1\omega)\underset{\cdot}{I}_1=\varepsilon^{j\theta_1}\underset{\cdot}{I}_1\sqrt{R_1^2+L_1^2\omega^2},$$

and

$$E = I_1\sqrt{R_1^2 + L_1^2\omega^2}.$$

E is ahead of I_1 in phase by the angle θ_1. In a similar way we have

$$\dot{E} = (R_2 + jL_2\omega)\dot{I}_2 = \varepsilon^{j\theta_2}\dot{I}_2\sqrt{R_2^2 + L_2^2\omega^2},$$

and

$$E = I_2\sqrt{R_2^2 + L_2^2\omega^2}.$$

The phase of I_2 is behind that of E by the angle θ_2. Combining the equations for $\dot{E}$, we obtain the relation between I_2 and I_1 as follows:

$$\dot{I}_2 = \varepsilon^{j(\theta_1 - \theta_2)}\dot{I}_1\sqrt{\frac{R_1^2 + L_1^2\omega^2}{R_2^2 + L_2^2\omega^2}} = \varepsilon^{j(\theta_1 - \theta_2)}\dot{I}_1 A,$$

where A is written for the expression $\sqrt{\dfrac{R_1^2 + L_1^2\omega^2}{R_2^2 + L_2^2\omega^2}}$,

and

$$I_2 = I_1\sqrt{\frac{R_1^2 + L_1^2\omega^2}{R_2^2 + L_2^2\omega^2}} = I_1 A.$$

I_2 is ahead of I_1 in phase by the angle $\theta_1 - \theta_2$. The whole current I, that is the current in the undivided portion of the circuit, is

$$\dot{I} = \dot{I}_1 + \dot{I}_2 = \dot{I}_1(1 + \varepsilon^{j(\theta_1 - \theta_2)}A).$$

Separating real and imaginary parts of $\dot{I}$, we have

$$\dot{I} = \dot{I}_1(1 + A\cos(\theta_1 - \theta_2) + jA\sin(\theta_1 - \theta_2)).$$

Writing

$$\theta_3 = \tan^{-1}\frac{A\sin(\theta_1 - \theta_2)}{1 + A\cos(\theta_1 - \theta_2)},$$

and deriving the values of $\sin\theta_3$ and $\cos\theta_3$, we obtain

$$\dot{I} = \dot{I}_1(\cos\theta_3 + j\sin\theta_3)\sqrt{1 + A^2 + 2A\cos(\theta_1 - \theta_2)},$$

or

$$\dot{I} = \varepsilon^{j\theta_3}\dot{I}_1\sqrt{1 + A^2 + 2A\cos(\theta_1 - \theta_2)},$$

and

$$I = I_1\sqrt{1 + A^2 + 2A\cos(\theta_1 - \theta_2)}.$$

The phase of I leads the phase of I_1 by the angle $\theta_1-\theta_3$. It is evident that precisely the same equations would have been reached if, instead of assuming I_1 to be known, we had assumed knowledge of E, I_2 or I. The equations would simply have been derived in a different order.

Example. Assume $I_1=100$ amperes, $R_1=5$ ohms, $L_1\omega=2.5$ ohms, $R_2=15$ ohms, $L_1\omega=15$ ohms, to find E, I_2 and I and the phase relations.

Answers. $E=559.0$, $I_2=26.352$, $I=125.28$, $\theta_1=\tan^{-1}\frac{1}{2}$, $\theta_2=\tan^{-1}1=45°$, $\theta_3=\tan^{-1}0.066708=3°\ 49'\ 18''$.

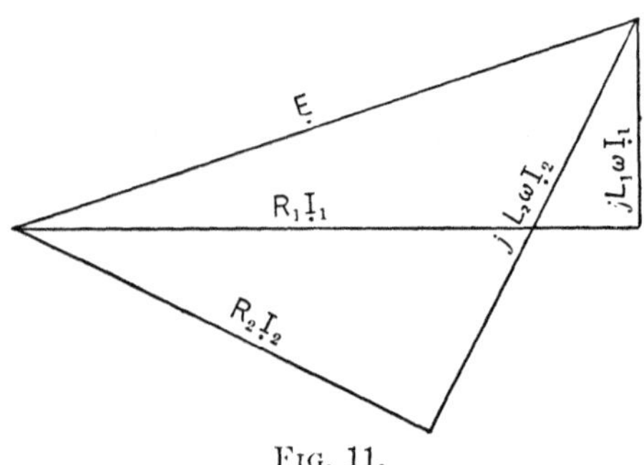

Fig. 11.

§ 24. The problem of a divided circuit is illustrated graphically in the diagrams (Figs. 11 and 12). The diagrams illustrate a problem in which the resistance of the first branch is three times the reactance, while in the second branch resistance and reactance are equal. From the former diagram (Fig. 11) the e.m.f. E may be determined in terms of R_1, $L_1\omega$, and I_2. After E is determined, the diagram furnishes the means of finding R_2I_2, and the value of I_2 may be determined. The latter diagram (Fig. 12) shows the relation of I_1, I_2 and I. In this I_1 is drawn parallel to R_1I_1 of the other diagram, and in the same way I_2 is parallel to R_2I_2. The diagonal represents I to the same scale as I_1 and I_2. If the diagrams are drawn carefully to scale they are an excellent check on the accuracy of the analytical solution, though it is evident that the analytical method must be more accurate if carried through without error.

In both the case of series circuits and that of divided circuits it is possible that the quantities to be added may differ in phase by large angles and the total (so called) of the e.m.f. or current, in the different cases respectively may be less than either component. This is the case in a series circuit including a motor and a generator, which

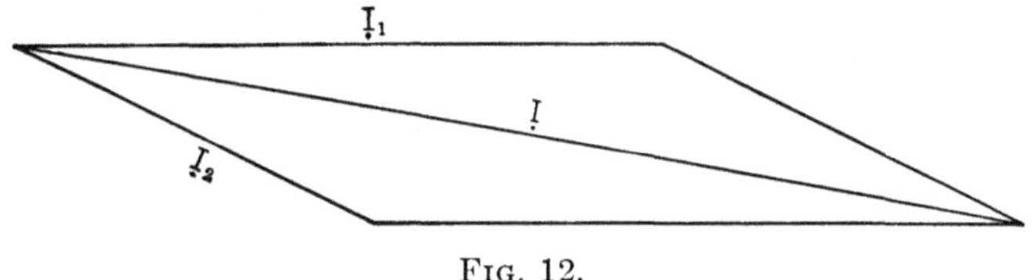

FIG. 12.

in practical cases are nearly opposite in phase, or with condensers and inductances in series. This is true also of currents in a divided circuit, one branch having inductance and the other capacity.

RESOLUTION INTO COMPONENTS

§ 25. Instead of indicating the e.m.f. and current as projections of uniform circular quantities expressed in magnitude and phase (the latter as a function of the time), we may express the circular quantities at the particular instant in terms of their real and imaginary components. Let E_1+jE_2 represent the circular quantity whose magnitude $\sqrt{E_1^2+E_2^2}$ is the effective value of the e.m.f. The instantaneous value of the e.m.f. is $\sqrt{2}E_1$, as explained elsewhere (§19). The expression I_1+jI_2 in the same way expresses the current (effective value $=\sqrt{I_1^2+I_2^2}$, and instantaneous value $\sqrt{2}I_1$). We then shall have

$$E_1+jE_2=\left(R+j\left(L\omega-\frac{1}{C\omega}\right)\right)(I_1+jI_2)$$

$$=RI_1-\left(L\omega-\frac{1}{C\omega}\right)I_2+j\left[RI_2+\left(L\omega-\frac{1}{C\omega}\right)I_1\right].$$

As the two circular quantities are equal in both magnitude and direction, we have the two equations,

$$E_1=RI_1-\left(L\omega-\frac{1}{C\omega}\right)I_2,$$

and

$$E_2=RI_2+\left(L\omega-\frac{1}{C\omega}\right)I_1.$$

The quantities E_1, E_2, I_1, I_2, may be positive or negative, or some may be positive and the others negative.

We also have

$$I_1+jI_2=\frac{E_1+jE_2}{R+j\left(L\omega-\frac{1}{C\omega}\right)}=\frac{(E_1+jE_2)\left(R-j\left(L\omega-\frac{1}{C\omega}\right)\right)}{R^2+\left(L\omega-\frac{1}{C\omega}\right)^2},$$

$$=\frac{E_1R+E_2\left(L\omega-\frac{1}{C\omega}\right)}{R^2+\left(L\omega-\frac{1}{C\omega}\right)^2}+j\frac{E_2R-E_1\left(L\omega-\frac{1}{C\omega}\right)}{R^2+\left(L\omega-\frac{1}{C\omega}\right)^2},$$

and this is equivalent to the two equations,

$$I_1=\frac{E_1R+E_2\left(L\omega-\frac{1}{C\omega}\right)}{R^2+\left(L\omega-\frac{1}{C\omega}\right)^2},$$

and

$$I_2=\frac{E_2R-E_1\left(L\omega-\frac{1}{C\omega}\right)}{R^2+\left(L\omega-\frac{1}{C\omega}\right)^2}.$$

Lastly we have

$$R+j\left(L\omega-\frac{1}{C\omega}\right)=\frac{E_1+jE_2}{I_1+jI_2}=\frac{(E_1+jE_2)(I_1-jI_2)}{I_1^2+I_2^2}$$

$$=\frac{E_1I_1+E_2I_2}{I_1^2+I_2^2}+j\frac{E_2I_1-E_1I_2}{I_1^2+I_2^2},$$

and from this it follows that

$$R=\frac{E_1I_1+E_2I_2}{I_1^2+I_2^2},$$

and

$$L\omega-\frac{1}{C\omega}=\frac{E_2I_1-E_1I_2}{I_1^2+I_2^2}.$$

Should the circuit be non-inductive, we shall have $L=0$ with corresponding changes in the formulæ. If there is no capacity in the circuit, we must not assume C equal to zero; for a condenser with zero capacity means an open circuit. We must instead simply remove the term $\frac{1}{C\omega}$. It is interesting to note that this comes to the same result as if the capacity had become infinite; in which case a finite charge (time integral of the current) would not cause an appreciable potential difference between condenser terminals. That is, the condenser will not interpose any direct or counter e.m.f. in the circuit.

§ 26. The problem of series circuits with more than one e.m.f., resistance, inductance, and capacity is to be solved by using ΣE_1, ΣE_2, ΣR, $\Sigma L\omega$, and $\Sigma\frac{1}{C\omega}$ in place of the single quantities.

The problem of divided circuits is treated in an analogous way to the divided circuit problem (§§ 23, 24), which we have already considered, by substituting $L_1\omega-\frac{1}{C_1\omega}$ in place of $L_1\omega$ for the first branch and substituting corresponding expressions for $L_2\omega$, etc., in the other branches.

USE OF A SYMMETRICAL PAIR OF TRIANGLES

§ 27. To represent the e.m.f. by a pair of uniform circular motions, in terms of the current, resistance, and reactance, as expressed by the equation

$$e=\frac{E}{2}(\varepsilon^{j\omega t}+\varepsilon^{-j\omega t})=\frac{I}{2}[(R+jL\omega)\varepsilon^{j(\omega t-\theta)}+(R-jL\omega)\varepsilon^{-j(\omega t-\theta)}],$$

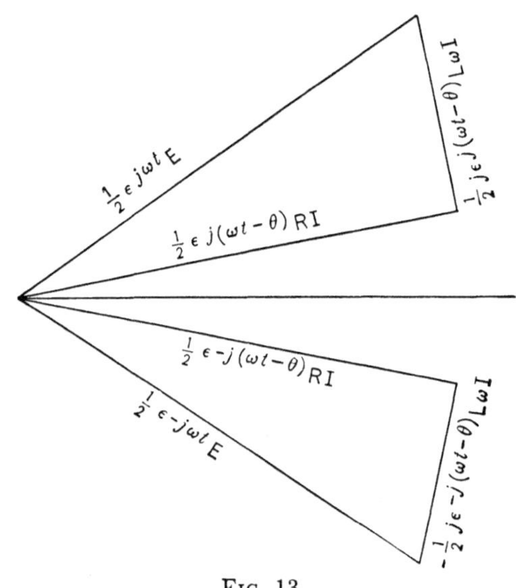

FIG. 13.

we may make use of two triangles revolving in opposite sense with angular velocities ω and always symmetrical with respect to the horizontal line (Fig. 13). While this mode of representing a simple harmonic quantity is complete and has no parts to be rejected, it is evidently too complicated for general acceptance and use by engineers.

CHAPTER V

PRODUCT OF TWO HARMONIC QUANTITIES

§ 28. Let us now consider the product of harmonic quantities and in particular the power of an electric circuit, the product of current and electromotive force. It will be seen in general that the product of two simple harmonic quantities is not *simple* harmonic, but *compound* harmonic. In the particular case of greatest interest to us, that is electric power, the product may be resolved into a constant plus a simple harmonic quantity of double frequency. It has been shown earlier that if two complex expressions be multiplied, the product will be complex. For example the product of

$$A = (\cos\theta + j\sin\theta)R \quad \text{and} \quad B = (\cos\phi + j\sin\phi)S$$

is

$$A \cdot B = (\cos(\theta + \phi) + j\sin(\theta + \phi))R \cdot S.$$

In general the rule of multiplication is as follows: The product has a magnitude equal to the product of the magnitudes of the factors, and makes an angle with the axis of reals equal to the sum of the angles made by the factors with that axis. If the factors are uniform circular in form and functions of the time t and angular velocities ω_1 and ω_2 as follows:

$$A = (\cos\omega_1 t + j\sin\omega_1 t)R,$$

$$B = (\cos\omega_2 t + j\sin\omega_2 t)S,$$

the product is

$$A \cdot B = (\cos\,(\omega_1+\omega_2)t + j \sin\,(\omega_1+\omega_2)t) R \cdot S,$$

a result which indicates that the product has a magnitude equal to the product of the magnitudes of the factors and an angular velocity equal to the sum of those of the factors. If ω_1 equals ω_2, the product has an angular velocity twice as great as the factors.

§ 29. For the sake of a *reductio ad absurdum* let us assume that the power of an electric circuit may be obtained from such a product of two uniform circular expressions for current and e.m.f. as follows:

$$\dot{I} = (\cos\,(\omega t - \theta) + j \sin\,(\omega t - \theta)) I,$$

$$\dot{E} = (\cos\,\omega t + j \sin\,\omega t) E,$$

and

$$\dot{E}\dot{I} = (\cos(2\omega t - \theta) + j \sin\,(2\omega t - \theta)) EI.$$

The product $\dot{E}\dot{I}$ is a uniform circular quantity of double frequency, as shown by the factor $(2\omega t - \theta)$, and has an average value zero for its projection on the axis of reals. This product evidently is not power, for the power of an electric circuit has in general an average value different from zero.

POWER IN SIMPLE CIRCUITS

§ 30. Let us now take the similar simple harmonic expressions for current and e.m.f. in terms of effective values of I and E:

$$i = \sqrt{2} I \cos\,(\omega t - \theta) = \frac{\sqrt{2}}{2} I (\varepsilon^{j(\omega t - \theta)} + \varepsilon^{-j(\omega t - \theta)}),$$

$$e = \sqrt{2} E \cos\,\omega t = \frac{\sqrt{2}}{2} E (\varepsilon^{j\omega t} + \varepsilon^{-j\omega t}).$$

By multiplication we obtain the power p, with average value P,

$$p=ei=2EI\cos\omega t\cdot\cos(\omega t-\theta)$$

$$=\frac{EI}{2}[\varepsilon^{j(2\omega t-\theta)}+\varepsilon^{-j(2\omega t-\theta)}+\varepsilon^{j\theta}+\varepsilon^{-j\theta}]$$

$$=EI[\cos\theta+\cos(2\omega t-\theta)]=P+EI\cos(2\omega t-\theta),$$

$$=P+\frac{P}{\cos\theta}\cos(2\omega t-\theta).$$

This expression shows that the instantaneous value of the power is equal to a constant P plus a simple harmonic quantity $\frac{P}{\cos\theta}\cos(2\omega t-\theta)$ of twice the frequency of the current and the e.m.f. This may be expressed in circular form provided the origin O' be taken eccentric to the circle.

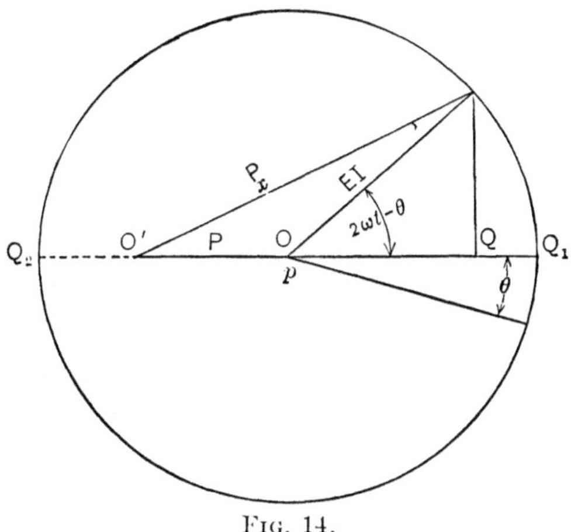

FIG. 14.

The diagram (Fig. 14) expresses the power in circular form. The instantaneous value, p, of the power is expressed by the distance and sense of $O'Q$. The maximum value of the power is $O'Q_1$, and the minimum (negative maximum) is $O'Q_2$. As P cannot be greater than EI, and may only equal EI when θ is zero, the point O' must not be exterior to the circle. The circular formula is as follows:

$$\underset{x}{P} = P + [E\dot{I}] = P + (\cos(2\omega t - \theta) + j\sin(2\omega t - \theta))EI,$$

where $\underset{x}{P}$ denotes an *eccentric* uniform circular quantity made up of a constant P and the *concentric* uniform circular quantity $[E\dot{I}]$.

§ 31. It is interesting to see how Dr. Steinmetz by introducing an arbitrary method of multiplication is able to obtain the average value of the power from the circular formulæ (concentric) for $\dot{E}$ and $\dot{I}$. He says[1]:

> "For the double frequency vector P, $j^2 = +1$, or 360° rotation and $j \times 1 = j$ and $1 \times j = -j$. That is, multiplication by j reverses the sign," . . .

Applying his rule we obtain the correct result as follows:

$$\dot{E} = (\cos \omega t + j \sin \omega t)E,$$

$$\dot{I} = (\cos(\omega t - \theta) + j\sin(\omega t - \theta))I,$$

$$\dot{E}\dot{I} = [\cos \omega t \cdot \cos(\omega t - \theta) + \sin \omega t \sin(\omega t - \theta)$$
$$+ j(\sin \omega t \cdot \cos(\omega t - \theta) - \cos \omega t \cdot \sin(\omega t - \theta))]EI,$$
$$= [\cos \theta + j \sin \theta]EI.$$

The real component of the product $\dot{E}\dot{I}$ is the power $EI \cos \theta$ (average value), the imaginary component is the so-called wattless component of the power $EI \sin \theta$ in magnitude.

It may be urged in objection to this method of obtaining power, that while the real part of the product is the average value of the power, we must take the whole mag-

[1] Steinmetz, *Alternating Current Phenomena*, 3d edition p. 151.

nitude of the factors for effective values of the e.m.f. and current, and the real parts of these factors must be multiplied by $\sqrt{2}$ to give the instantaneous values of these quantities.

§ 32. Another arbitrary method of combining e.m.f. and current to obtain power is as follows: Let the e.m.f. and current be represented in effective value by the magnitudes of the complex quantities,

$$\underset{\cdot}{E} = E_1 + jE_2 = (\cos \omega t + j \sin \omega t)E,$$

$$\underset{\cdot}{I} = I_1 + jI_2 = (\cos (\omega t - \theta) + j \sin (\omega t - \theta))I.$$

We know from previous proof that the average value of the power P is $P = EI \cos \theta$, also that $\frac{E_1}{E} = \cos \omega t$, $\frac{E_2}{E} = \sin \omega t$, $\frac{I_1}{I} = \cos (\omega t - \theta)$, $\frac{I_2}{I} = \sin (\omega t - \theta)$ and that

$$\cos \theta = \cos (\omega t) \cos (\omega t - \theta) + \sin \omega t \cdot \sin (\omega t - \theta).$$

It therefore follows that

$$E_1 I_1 + E_2 I_2 = EI \cos \theta \text{ (a constant).}$$

As $EI \cos \theta$ has a constant value, although E_1, E_2, I_1 and I_2 are all variables, it is evident that we shall obtain the correct result for average value of the power if we take the values of these variables at any one time.

We therefore have as an arbitrary rule to obtain the average value of the power: multiply the real parts of e.m.f. and current, and the imaginary parts, ignoring the j^2 and add the products.

Thus

$$\underset{\cdot}{E} = E_1 + jE_2,$$

$$\underset{\cdot}{I} = I_1 + jI_2,$$

$$P = E_1 I_1 + E_2 I_2.$$

If a minus sign is expressed in one of the factors it must not be ignored. For example if we have

$$\dot{E}=E_1-jE_2,$$

$$\dot{I}=I_1+jI_2,$$

then

$$P=E_1I_1-E_2I_2.$$

If we have a circuit including a motor whose e.m.f. in general opposes the current, we have for example

$$\dot{E}=-E_1+jE_2,$$

$$\dot{I}=I_1+jI_2,$$

$$P=-E_1I_1+E_2I_2.$$

If we have two minus signs they must both be regarded, for example

$$\dot{E}=E_1-jE_2,$$

$$\dot{I}=I_1-jI_2,$$

$$P=E_1I_1+E_2I_2,$$

etc.

It is evident that the above examples are not examples of real multiplication of current and e.m.f. They are merely the expression of a rule for finding average power where the e.m.f. and current are known in magnitude and phase relation.

§ 33. The above process is not easily reversed; for if

$$P=E_1I_1+E_2I_2,$$

and supposing I_1 and I_2 known, it is evident that there are an indefinite number of values of E_1 (with corresponding values of E_2) which will satisfy the equation. The correct values of E_1 and E_2 can only be found when more data are given. It suffices when θ is known. We have had the known relation $P=EI \cos \theta$, and it may be shown that

$$E_1=P\frac{I_1-I_2 \tan \theta}{I_1^1+I_2^2},$$

and

$$E_2=P\frac{I_2+I_1 \tan \theta}{I_1^2+I_2^2}.$$

In these formulæ θ is the angle of lag of the current behind the e.m.f. If θ is taken as the angle of lead, the terms involving $\tan \theta$ must be altered accordingly.

It is on the whole more satisfactory to use another method and find directly from the relations,

$$P=EI \cos \theta,$$

and

$$\underset{\cdot}{I}=I(\cos (\omega t-\theta)+j \sin (\omega t-\theta)),$$

the result,

$$\underset{\cdot}{E}=\frac{P}{I \cos \theta} (\cos \omega t+j \sin \omega t).$$

POWER IN CIRCUITS OF MORE THAN ONE PHASE

§ 34. The power in circuits of more than one phase may be obtained by simple addition, whether we are dealing with instantaneous or average values. An interesting case is that of balanced circuits having two, three or more phases; for in every case of circuits of more than one phase the power is constant during the whole cycle if all the currents and e.m.f.'s of the various phases have equal effective values respectively, and if the phase differences from each to the next equals $\frac{360°}{n}$, where n is the number of phases. The two-phase circuit follows the same rule, though it does not come under the above statement of phase difference; for in it we have one interval of 90° and the next 270°.

BALANCED TWO-PHASE CIRCUIT

§ 35. Let us take the balanced two-phase circuit first. It has been shown that the power at any instant in an alternating current circuit, with simple harmonic e.m.f. and current is

$$p=EI(\cos\theta+\cos(2\omega t-\theta)).$$

If the current and e.m.f. in the second branch of the circuit have the same effective values as in the first, but a phase difference of $\frac{\pi}{2}$, the variable part of the power will have a phase difference of π, because power is a *double* frequency variable. We therefore shall have for p_1 and p_2, the power of the two phases,

$$p_1=EI(\cos\theta+\cos(2\omega t-\theta)),$$

$$p_2=EI(\cos\theta-\cos(2\omega t-\theta)),$$

and

$$p=p_1+p_2=2EI\cos\theta=P, \quad \text{(a constant)}.$$

BALANCED THREE-PHASE CIRCUIT

§ 36. For a balanced three-phase circuit we shall have, if the phase intervals are $120°=\frac{2\pi}{3}$,

$$p_1=EI(\cos\theta+\cos(2\omega t-\theta)),$$

$$p_2=EI(\cos\theta+\cos(2\omega t-\theta+\tfrac{4}{3}\pi)) \\ =EI(\cos\theta+\cos(2\omega t-\theta-\tfrac{2}{3}\pi)),$$

$$p_3=EI(\cos\theta+\cos(2\omega t-\theta+\tfrac{8}{3}\pi)) \\ =EI(\cos\theta+\cos(2\omega t-\theta+\tfrac{2}{3}\pi)),$$

and

$$p=p_1+p_2+p_3=3EI\cos\theta=P.$$

It is evident that the variable parts of p annul one another, for they may be represented as the projections of the three sides of an equilateral triangle, each side equal to EI in magnitude and the first side making an angle with the line on which they are projected equal to $2\omega t-\theta$. It is well known that the sum of the projections of the sides of any closed polygon is zero. This relation is illustrated by the diagram (Fig. 15).

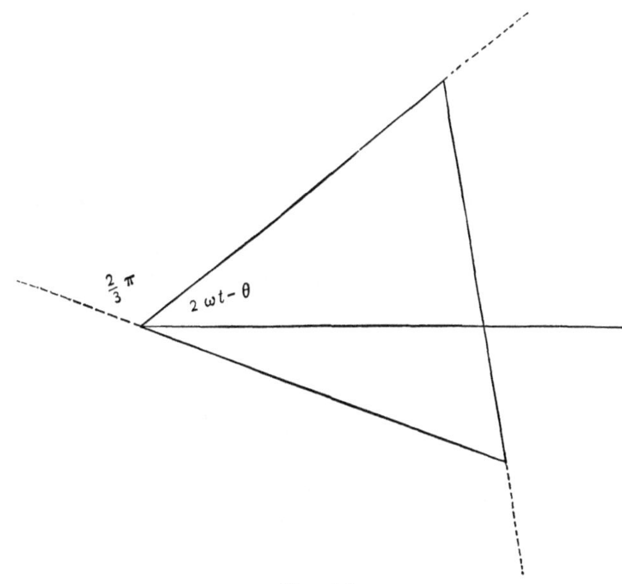

Fig. 15.

BALANCED FOUR-PHASE CIRCUIT

§ 37. The four-phase case is, in a similar way, shown to have constant power. Or it may be looked upon as two pairs of balanced two-phase, each pair as shown above having constant power. Therefore the sum of all four has constant power.

BALANCED SIX-PHASE CIRCUIT

§ 38. The six-phase case is evidently a case of two balanced three-phase systems. The variable part may be represented by the projections of the sides of an equilateral triangle each being taken twice. The power of the parts and the whole are

$$p_1 = EI(\cos\theta + \cos(2\omega t - \theta)),$$
$$p_2 = EI(\cos\theta + \cos(2\omega t - \theta + \tfrac{2}{3}\pi)),$$
$$p_3 = EI(\cos\theta + \cos(2\omega t - \theta + \tfrac{4}{3}\pi)),$$
$$p_4 = EI(\cos\theta + \cos(2\omega t - \theta + 2\pi)) = p_1,$$
$$p_5 = EI(\cos\theta + \cos(2\omega t - \theta + \tfrac{8}{3}\pi)) = p_2,$$
$$p_6 = EI(\cos\theta + \cos(2\omega t - \theta + \tfrac{10}{3}\pi)) = p_3,$$
$$p = p_1 + p_2 + p_3 + p_4 + p_5 + p_6 = 6EI\cos\theta = P.$$

BALANCED POLYPHASE CIRCUITS IN GENERAL

§ 39. If there are an odd number of phases, more than three, the variable part of the power is represented by

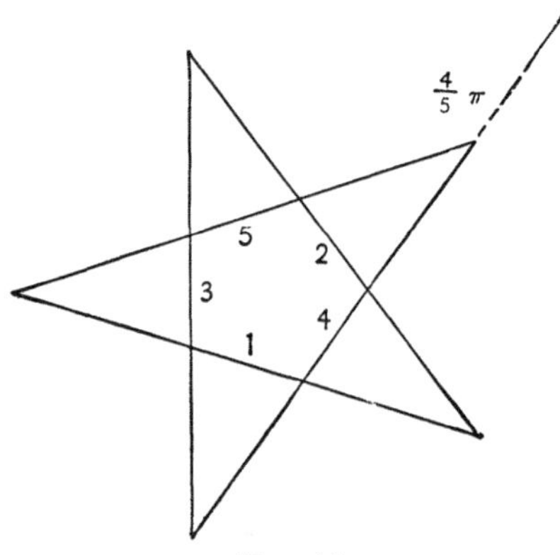

FIG. 16.

the projections of the lines of a star-shaped diagram which is in all cases a closed figure with a total of zero for the projections. The five-phase case will suffice for illustration (Fig. 16). For five phases the progressive interval is $\frac{2}{5}\pi = 72°$ for current and e.m.f. and $\frac{4}{5}\pi = 144°$ for the variable part of the power. The power of the various phases and the total are as follows:

$$p_1 = EI(\cos\theta + \cos(2\omega t - \theta)),$$

$$p_2 = EI(\cos\theta + \cos(2\omega t - \theta + \tfrac{4}{5}\pi)),$$

$$p_3 = EI(\cos\theta + \cos(2\omega t - \theta + \tfrac{8}{5}\pi)) \\ = EI(\cos\theta + \cos(2\omega t - \theta - \tfrac{2}{5}\pi)),$$

$$p_4 = EI(\cos\theta + \cos(2\omega t - \theta + \tfrac{12}{5}\pi)) \\ = EI(\cos\theta + \cos(2\omega t - \theta + \tfrac{2}{5}\pi)),$$

$$p_5 = EI(\cos\theta + \cos(2\omega t - \theta + \tfrac{16}{5}\pi)) \\ = EI(\cos\theta + \cos(2\omega t - \theta - \tfrac{4}{5}\pi)),$$

$$p = p_1 + p_2 + p_3 + p_4 + p_5 = 5EI\cos\theta = P.$$

The eccentric circular diagram for the power of a balanced polyphase system of n phases reduces evidently

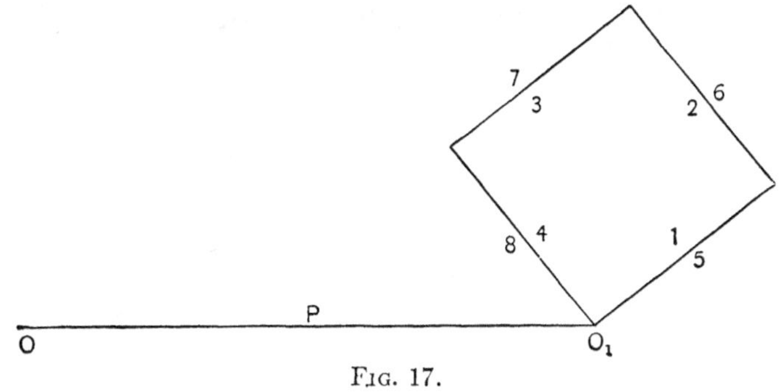

FIG. 17.

to a horizontal line of length $nEI\cos\theta$ and a closed regular polygon of $\frac{n}{2}$ sides with each side taken twice if n is even, or a regular star of n sides if n is odd. The diagram (Fig. 17) illustrates an eight-phase balanced system, in which the instantaneous power is a constant quantity P. The variable part of the power of each separate phase is the projection of the corresponding side of the square, each being taken twice as indicated.

In the same way the diagram (Fig. 18) represents the power of a five-phase balanced system. As before, the

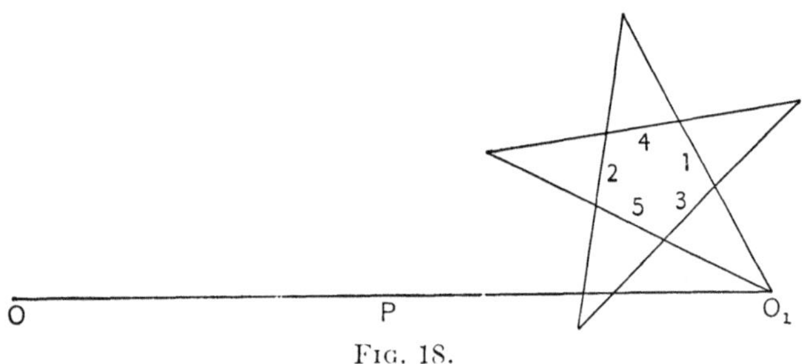

Fig. 18.

instantaneous power is constant for the system, the variable part of the power of the separate phases being represented by the projections of the lines of the five-pointed star.

UNBALANCED POLYPHASE CIRCUITS

§ 40. While in general we expect the power to be constant only in balanced symmetrical polyphase (or two- or

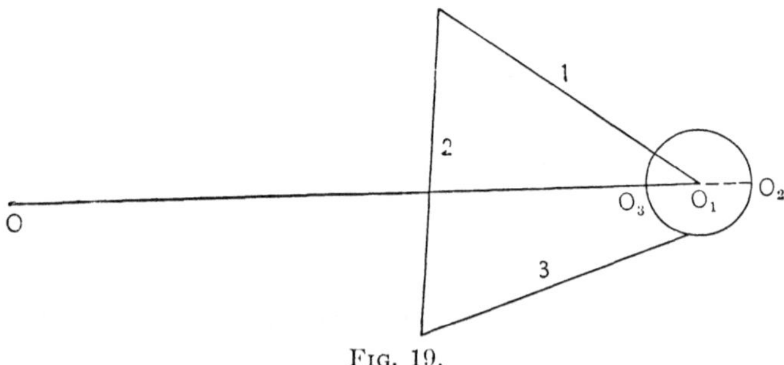

Fig. 19.

three-phase) systems, it is evident that any system will have constant power if the eccentric circular diagram gives a closed figure for the variable parts of the separate phases. As a rule unbalanced polyphase systems do not give a closed figure for the variable parts of the eccentric circular diagram.

The diagram (Fig. 19) illustrates the power of an unbalanced three-phase system for which P, the average value of the power, is represented by OO_1.

The maximum value of p is OO_2 and the minimum is OO_3.

CHAPTER VI

NON-HARMONIC CURRENTS

§ 41. The method of revolving vectors has interesting applications to cases of currents which are not harmonic in the strict sense of the word. The cases which we shall here investigate are, *first,* the oscillatory discharge of a condenser, *second,* its non-oscillatory discharge and *third,* the current following the closing of a circuit in which the e.m.f. is simple harmonic. In this last case it is well known that the current is not harmonic, but as time goes on approaches more and more nearly to harmonic values.

OSCILLATORY DISCHARGE OF A CONDENSER

§ 42. First let us consider the oscillatory discharge of a condenser. Let the capacity of the condenser be represented by C, the e.m.f. to which it is charged by E_0 when the circuit is about to be closed, and by e at later times. Let the current be indicated by i, the resistance of the circuit by R and the inductance by L. It will be assumed that C, R, and L are all constants. We shall have as the form of Ohm's law applicable to variable conditions

$$e = Ri + L\frac{di}{dt} = E_0 - \frac{\int i dt}{C},$$

or

$$\int i dt + RCi + LC\frac{di}{dt} = CE_0.$$

Differentiating this expression we have, after rearranging the terms

$$LC\frac{d^2i}{dt^2}+RC\frac{di}{dt}+i=0.$$

This equation has two solutions, each of the form

$$i=K\varepsilon^{-\alpha t},$$

where K and α are constants to be determined. The constant K depends on E_0, as will be shown later, and cannot be found from the differential equation in its latter form. The values of α are, however, to be found; for substituting $i=K\varepsilon^{-\alpha t}$, we obtain

$$(\alpha^2LC-\alpha RC+1)K\varepsilon^{-\alpha t}=0.$$

We shall assume that the current i, which equals $K\varepsilon^{-\alpha t}$, is not zero in general. We therefore have

$$\alpha^2LC-\alpha RC+1=0,$$

$$\alpha^2-\alpha\frac{R}{L}+\frac{1}{LC}=0,$$

$$\alpha^2-\alpha\frac{R}{L}+\frac{R^2}{4L^2}=\frac{R^2}{4L^2}-\frac{1}{LC},$$

and

$$\alpha=\frac{R}{2L}+\sqrt{\frac{R^2}{4L^2}-\frac{1}{LC}}=\frac{R}{2L}\left(1+\sqrt{1-\frac{4L}{R^2C}}\right)=\alpha_1,$$

or

$$\alpha=\frac{R}{2L}-\sqrt{\frac{R^2}{4L^2}-\frac{1}{LC}}=\frac{R}{2L}\left(1-\sqrt{1-\frac{4L}{R^2C}}\right)=\alpha_2.$$

The general solution for i may include both values for α and different values for K which we shall designate by K_1 and K_2. The equation for the current is then

$$i=K_1\varepsilon^{-\alpha_1t}+K_2\varepsilon^{-\alpha_2t}.$$

As it is evident that at the time t of closing the circuit, $t=0$ and $i=0$, we shall have therefore

$$O=K_1+K_2, \quad \text{or} \quad K_1=-K_2.$$

and

$$i=K_1(\varepsilon^{-\alpha_1 t}-\varepsilon^{-\alpha_2 t}).$$

If α_1 and α_2 are real quantities, an investigation of this equation will show that the current will start at a value zero, rise to a maximum, and fall off later to smaller and smaller values without reversal of sign, and become zero only after an infinite time has elapsed. This condition is expressed by the inequality $R^2C>4L$. If on the other hand $R^2C<4L$, it is evident that α_1 and α_2 are complex quantities as follows:

$$\alpha_1=\frac{R}{2L}+j\frac{R}{2L}\sqrt{\frac{4L}{R^2C}-1}=\frac{R}{2L}+j\beta,$$

$$\alpha_2=\frac{R}{2L}-j\frac{R}{2L}\sqrt{\frac{4L}{R^2C}-1}=\frac{R}{2L}-j\beta,$$

where we have written

$$\frac{R}{2L}\sqrt{\frac{4L}{R^2C}-1}=\beta.$$

Converting the exponentials with imaginary indices into sine and cosine terms, and remembering that in this particular case the current must be zero when t is zero, we have

$$i=K\varepsilon^{-\frac{R}{2L}t}\sin\beta t=K\varepsilon^{-\frac{R}{2L}t}\sin\left(\frac{Rt}{2L}\sqrt{\frac{4L}{R^2C}-1}\right).$$

The period of the oscillation of the discharge (or current) is

$$T=\frac{4\pi LC}{\sqrt{4LC-R^2C^2}}.$$

If R^2 is small in comparison with $\frac{4L}{C}$, the period is

$$T=2\pi\sqrt{LC} \quad \text{(approximately)}.$$

With larger values of R the period is increased until when R equals $2\sqrt{\frac{L}{C}}$, the period becomes infinite. If R^2 is greater than $\frac{4L}{C}$ the discharge is aperiodic (without a period), corresponding to the earlier formula.

Let us now investigate e, the e.m.f. which equals the potential difference between condenser terminals. We have the relation

$$e=Ri+L\frac{di}{dt}.$$

Substituting the value of i as given above, we have

$$e=K\varepsilon^{-\frac{R}{2L}t}\left[R\sin\beta t+L\beta\cos\beta t-\frac{R}{2}\sin\beta t\right]$$

$$=\frac{KR}{2}\varepsilon^{-\frac{R}{2L}t}\left[\sin\beta t+\sqrt{\frac{4L}{R^2C}-1}\cos\beta t\right].$$

Writing, to simplify the expression, $\tan\delta=\sqrt{\frac{4L}{R^2C}-1}$, we have

$$e=K\sqrt{\frac{L}{C}}\varepsilon^{-\frac{R}{2L}t}\sin(\beta t+\delta).$$

As on closing the circuit (*i.e.*, $t=0$) we had $E_0=e$, we obtain as the value of the constant K,

$$K=\frac{2E_0\sqrt{C}}{\sqrt{4L-R^2C}},$$

and

$$e=\frac{2E_0\sqrt{L}}{\sqrt{4L-R^2C}}\varepsilon^{-\frac{R}{2L}t}\sin\left(\frac{Rt}{2L}\sqrt{\frac{4L}{R^2C}-1}+\tan^{-1}\sqrt{\frac{4L}{R^2C}-1}\right),$$

and

$$i=\frac{2E_0\sqrt{C}}{\sqrt{4L-R^2C}}\varepsilon^{-\frac{R}{2L}t}\sin\left(\frac{Rt}{2L}\sqrt{\frac{4L}{R^2C}-1}\right).$$

If R^2 is small in comparison with $\frac{4L}{C}$, the current for early oscillations is

$$i=E_0\sqrt{\frac{C}{L}}\sin\frac{t}{\sqrt{LC}} \quad \text{(approximately)}.$$

The maximum value of the current (complete expression) occurs in less than one-quarter of a period after closing the circuit, at a time when

$$\sin\left(\frac{Rt}{2L}\sqrt{\frac{4L}{R^2C}-1}\right)=\sqrt{1-\frac{R^2C}{4L}},$$

as may be shown by putting $\frac{di}{dt}$ equal to zero. The minimum value of i occurs half a period later, when the sine equals $-\sqrt{1-\frac{R^2C}{4L}}$, a result also of the same equation, $\frac{di}{dt}=0$.

Using the notation of rotating vectors we may express e and i as the real part of two exponential spirals $\underset{s}{E}$ and $\underset{s}{I}$ as follows:

$$\underset{s}{E}=\frac{2E_0\sqrt{L}}{\sqrt{4L-R^2C}}\varepsilon^{-\frac{R}{2L}t+j\left(\beta t+\delta-\frac{\pi}{2}\right)},$$

$$\underset{s}{I}=\frac{2E_0\sqrt{C}}{\sqrt{4L-R^2C}}\varepsilon^{-\frac{R}{2L}t+j\left(\beta t-\frac{\pi}{2}\right)}.$$

The charge of the condenser is evidently

$$q=eC=\frac{2E_0C\sqrt{L}}{\sqrt{4L-R^2C}}\varepsilon^{-\frac{Rt}{2L}}\sin\left(\frac{Rt}{2L}\sqrt{\frac{4L}{R^2C}-1}+\tan^{-1}\sqrt{\frac{4L}{R^2C}-1}\right).$$

This may be written as the real part of the exponential spiral

$$\underset{s}{Q}=\frac{2E_0C\sqrt{L}}{\sqrt{4L-R^2C}}\varepsilon^{-\frac{Rt}{2L}+j\left(\beta t+\delta-\frac{\pi}{2}\right)}.$$

§ 43. The results for i, e, and q might bave been deduced in the form of the difference of two exponential spirals, directly from the solution,

$$i=K_1(\varepsilon^{-\alpha_1 t}-\varepsilon^{-\alpha_2 t});$$

for if α_1 and α_2 are complex both $K_1\varepsilon^{-\alpha_1 t}$ and $K_1\varepsilon^{-\alpha_2 t}$ are exponential spirals. If the result is to be real at every instant, it is evident that the imaginary parts of both spirals must be equal, while the real parts must be equal in magnitude but of opposite sign. It is evident that both spirals must start for $t=0$ at a point for which the real values are zero. To satisfy this condition K_1 must be a pure imaginary. As we shall prefer to keep the constant real, we may reach the same result by changing the phase of the spirals by an angle $\frac{\pi}{2}$. Remembering that $\varepsilon^{j\frac{\pi}{2}}=j$ and $\varepsilon^{-j\frac{\pi}{2}}=-j$, we may write i as the *sum* of two exponential spirals, or

$$i=K\left(\varepsilon^{-\alpha_1 t+j\frac{\pi}{2}}+\varepsilon^{-\alpha_2 t-j\frac{\pi}{2}}\right),$$

where K is the magnitude of the pure imaginary $\cdot K_1$ as explained above.

Substituting the value of i in the equation for e,

$$e = Ri + L\frac{di}{dt},$$

we obtain

$$e = K\left[(R-\alpha_1 L)\varepsilon^{-\alpha_1 t + j\frac{\pi}{2}} + (R-\alpha_2 L)\varepsilon^{-\alpha_2 t - j\frac{\pi}{2}}\right],$$

or

$$e = K\left[\left(\frac{R}{2} - j\frac{R}{2}\sqrt{\frac{4L}{R^2C}-1}\right)\varepsilon^{-\alpha_1 t + j\frac{\pi}{2}} + \left(\frac{R}{2} + j\frac{R}{2}\sqrt{\frac{4L}{R^2C}-1}\right)\varepsilon^{-\alpha_2 t - j\frac{\pi}{2}}\right].$$

Writing as before $\tan\delta = \sqrt{\frac{4L}{R^2C}-1}$, $\sin\delta = \sqrt{1-\frac{R^2C}{4L}}$, etc., and remembering that $1 \pm j\sqrt{\frac{4L}{R^2C}-1} = \sqrt{\frac{4L}{R^2C}}\varepsilon^{\pm j\delta}$, we have after rearranging the terms,

$$e = K\sqrt{\frac{L}{C}}\varepsilon^{-\frac{Rt}{2L}}\left[\varepsilon^{j\left(\beta t - \frac{\pi}{2} + \delta\right)} + \varepsilon^{-j\left(\beta t - \frac{\pi}{2} + \delta\right)}\right].$$

At the time of closing the circuit (i.e., $t=0$), we had $e=E_0$, therefore

$$E_0 = K\sqrt{\frac{L}{C}}\left[\varepsilon^{j\left(\delta - \frac{\pi}{2}\right)} + \varepsilon^{-j\left(\delta - \frac{\pi}{2}\right)}\right],$$

or

$$E_0 = 2K\sqrt{\frac{L}{C}}\cos\left(\delta - \frac{\pi}{2}\right) = 2K\sqrt{\frac{L}{C}}\sin\delta = K\sqrt{\frac{4L-R^2C}{C}},$$

and

$$K = \frac{E_0\sqrt{C}}{\sqrt{4L-R^2C}}.$$

Substituting the value of K in the earlier formulæ, we have

$$i=\frac{E_0\sqrt{C}}{\sqrt{4L-R^2C}}\varepsilon^{-\frac{Rt}{2L}}\left[\varepsilon^{j\left(\frac{Rt}{2L}\sqrt{\frac{4L}{R^2C}-1}-\frac{\pi}{2}\right)}+\varepsilon^{-j\left(\frac{Rt}{2L}\sqrt{\frac{4L}{R^2C}-1}-\frac{\pi}{2}\right)}\right]$$

$$=\frac{2E_0\sqrt{C}}{\sqrt{4L-R^2C}}\varepsilon^{-\frac{Rt}{2L}}\sin\left(\frac{Rt}{2L}\sqrt{\frac{4L}{R^2C}-1}\right),$$

and

$$e=\frac{E_0\sqrt{L}}{\sqrt{4L-R^2C}}\varepsilon^{-\frac{Rt}{2L}}\left[\varepsilon^{j\left(\frac{Rt}{2L}\sqrt{\frac{4L}{R^2C}-1}-\frac{\pi}{2}+\delta\right)}\right.$$

$$\left.+\varepsilon^{-j\left(\frac{Rt}{2L}\sqrt{\frac{4L}{R^2C}-1}-\frac{\pi}{2}+\delta\right)}\right]$$

$$=\frac{2E_0\sqrt{L}}{\sqrt{4L-R^2C}}\varepsilon^{-\frac{Rt}{2L}}\sin\left(\frac{Rt}{2L}\sqrt{\frac{4L}{R^2C}-1}+\tan^{-1}\sqrt{\frac{4L}{R^2C}-1}\right).$$

The charge of the condenser has a similar formula, derived from the relation, $q=eC$. It is unnecessary to write it out in full.

The exponential spirals for each case may be plotted in polar coordinates, and will be seen to be in all respects equal except that one is right handed and the other left handed. The epoch angles for the starting points are indicated in the exponent.

NON-OSCILLATORY DISCHARGE OF A CONDENSER

§ 44. It has been shown above, §§ 42 and 43, that the condition, $R^2C<4L$, corresponds to an oscillatory discharge of the condenser. If, on the other hand, we have either $R^2C>4L$ or $R^2C=4L$, then the discharge will be non-oscillatory.

The general solution

$$i=K(\varepsilon^{-\alpha_1 t}-\varepsilon^{-\alpha_2 t}),$$

in which α_1 and α_2 are real, corresponds to the condition $R^2C>4L$; but becomes indeterminate if $R^2C=4L$, or $\alpha_1=\alpha_2$. In this latter case we must resort to the particular solution,

$$i=(K_1+K_2t)\,\varepsilon^{-\alpha t}.$$

First, let us consider the general solution. At the time of closing the circuit, we have

$$E_0=L\left[\frac{di}{dt}\right]_{t=0}=LK(\alpha_2-\alpha_1),$$

or

$$K=\frac{E_0}{L(\alpha_2-\alpha_1)}.$$

Substituting the values of α_1 and α_2, § 42, we have

$$i=\frac{2RCE_0}{\sqrt{R^2C^2-4LC}}\varepsilon^{-\frac{Rt}{2L}}\left(\varepsilon^{\frac{Rt}{2L}\sqrt{1-\frac{4L}{R^2C}}}-\varepsilon^{-\frac{Rt}{2L}\sqrt{1-\frac{4L}{R^2C}}}\right).$$

If $R^2C=4L$, we have from the particular solution

$$K_1=0 \qquad \text{and} \qquad K_2=\frac{E_0}{L},$$

giving as a final result

$$i=\frac{E_0t}{L}\varepsilon^{-\frac{Rt}{2L}}.$$

PHENOMENA OBSERVED ON CLOSING THE CIRCUIT, STARTING TERM

§ 45. Let us now consider the expression for the current in a circuit which has just been closed, the e.m.f. being simple harmonic. As is well known, the current will not become harmonic at once, even though the electromotive force is precisely harmonic. The divergence from harmonic values is expressed in the equation for the current by a so-called "starting term." Let us consider a circuit of constant resistance R, inductance L, capacity C (all in series) and a simple harmonic e.m.f. e, equal to $E\cos\omega t$. We have as the expression for Ohm's law extended to variable e.m.f.

$$e=E\cos\omega t=Ri+L\frac{di}{dt}+\frac{1}{C}\int idt=\frac{E}{2}(\varepsilon^{j\omega t}+\varepsilon^{-j\omega t}).$$

It is well known that in such a circuit the current eventually will follow a simple harmonic law. Indicating the starting term by ψI, ψ being a function of the time t, and I being the maximum value of the current after the harmonic condition is reached, we have

$$i=\frac{I}{2}(\varepsilon^{j(\omega t-\theta)}+\varepsilon^{-j(\omega t-\theta)})-\psi I.$$

Let us substitute the value of i in the previous equation. We obtain

$$e=\frac{E}{2}(\varepsilon^{j\omega t}+\varepsilon^{-j\omega t})=\frac{I}{2}\left[\left(R+jL\omega-j\frac{1}{C\omega}\right)\varepsilon^{j(\omega t-\theta)}\right.$$
$$\left.+\left(R-jL\omega+j\frac{1}{C\omega}\right)\varepsilon^{-j(\omega t-\theta)}\right]-\left[R\psi+L\frac{d\psi}{dt}+\int\frac{\psi dt}{C}\right]I.$$

The constant of integration is reserved for the last term. As no particular values have been assumed for I or θ, it is evident that if we write $\tan\theta=\dfrac{L\omega-\dfrac{1}{C\omega}}{R}$ and $E^2=I^2\left(R^2+\left(L\omega-\dfrac{1}{C\omega}\right)^2\right)$, we shall have

$$\frac{E}{2}(\varepsilon^{j\omega t}+\varepsilon^{-j\omega t})$$
$$=\frac{I}{2}\left[\left(R+jL\omega-j\frac{1}{C\omega}\right)\varepsilon^{j(\omega t-\theta)}+\left(R-jL\omega+j\frac{1}{C\omega}\right)\varepsilon^{-j(\omega t-\theta)}\right],$$

and therefore

$$R\psi+L\frac{d\psi}{dt}+\int\frac{\psi dt}{C}=0.$$

Differentiating the last equation and dividing by L, we have

$$\frac{d^2\psi}{dt^2}+\frac{R}{L}\frac{d\psi}{dt}+\frac{\psi}{CL}=0.$$

We have already found the solution of an equation of this form (§ 42). If $4L>R^2C$ we may write at once

$$\psi=K\varepsilon^{-\frac{Rt}{2L}}\left[\varepsilon^{j\left(\frac{Rt}{2L}\sqrt{\frac{4L}{R^2C}-1}+\gamma\right)}+\varepsilon^{-j\left(\frac{Rt}{2L}\sqrt{\frac{4L}{R^2C}-1}+\gamma\right)}\right],$$

where K and γ are both real quantities. This is equivalent to dropping the γ from the expression and giving the exponentials in the bracket different factors K_1 and K_2. If the latter mode of expression were used, K_1 and K_2 would in general be found to be complex constants.

If $R^2C>4L$ the solution for ψ takes the form

$$\psi=K_1\varepsilon^{-\frac{Rt}{2L}\left(1+\sqrt{1-\frac{4L}{R^2C}}\right)}+K_2\varepsilon^{-\frac{Rt}{2L}\left(1-\sqrt{1-\frac{4L}{R^2C}}\right)},$$

which we shall discuss later in § 48.

§ 46. To determine the values of K and γ, we must know the condition of the circuit at the time it is closed. As we have already assumed in our formula for the electromotive force that e is at its largest value when $t=0$, or any number of complete periods later, we cannot in fairness assume that the time of closing the circuit is necessarily the same. Let us then take the time of closing the circuit to be t_0. Let us assume that the condenser was already charged to a potential E_0 when put in circuit. We therefore have the conditions that

$$\left.\begin{aligned} i&=0 \quad \text{and} \\ L\frac{di}{dt}&=E\cos\omega t_0-E_0 \end{aligned}\right\}\ \text{at the time, } t_0\text{, of closing the circuit.}$$

Indicating by ψ_0 the value of ψ when $t=t_0$, we evidently have from the relations above,

$$\psi_0=\cos(\omega t_0-\theta)$$

and

$$E\cos\omega t_0-E_0=L\omega I\cos\left(\omega t_0-\theta+\frac{\pi}{2}\right)-LI\left[\frac{d\psi}{dt}\right]_{t=t_0},$$

or

$$\left[\frac{d\psi}{dt}\right]_{t=t_0}=-\frac{E\cos\omega t_0-E_0}{LI}-\omega\sin(\omega t_0-\theta).$$

Writing, as in § 42,

$$\beta=\frac{R}{2L}\sqrt{\frac{4L}{R^2C}-1},$$

the equation for the starting term, § 45, becomes

$$\psi=K\varepsilon^{-\frac{Rt}{2L}}[\varepsilon^{j(\beta t+\gamma)}+\varepsilon^{-j(\beta t+\gamma)}]=2K\varepsilon^{-\frac{Rt}{2L}}\cos(\beta t+\gamma);$$

and when $t=t_0$ we have

$$\psi_0=2K\varepsilon^{-\frac{Rt_0}{2L}}\cos(\beta t_0+\gamma)=\cos(\omega t_0-\theta).$$

We have also

$$\left[\frac{d\psi}{dt}\right]_{t=t_0}=-2K^{-\frac{Rt_0}{2L}}\left[\frac{R}{2L}\cos(\beta t_0+\gamma)+\beta\sin(\beta t_0+\gamma)\right]$$

$$\doteq-\frac{E\cos\omega t_0-E_0}{LI}-\omega\sin(\omega t_0-\theta).$$

From these equations we obtain

$$\tan(\beta t_0+\gamma)=\frac{E\cos\omega t_0-E_0+L\omega I\sin(\omega t_0-\theta)}{\beta LI\cos(\omega t_0-\theta)}-\frac{R}{2\beta L},$$

and

$$\gamma=\tan^{-1}\left[\frac{E\cos\omega t_0-E_0+L\omega I\sin(\omega t_0-\theta)}{\beta LI\cos(\omega t_0-\theta)}-\frac{R}{2\beta L}\right]-\beta t_0$$

$$=\tan^{-1}\left[\frac{2C}{\sqrt{4LC-R^2C^2}}\left(\frac{E\cos\omega t_0-E_0+L\omega I\sin(\omega t_0-\theta)}{I\cos(\omega t_0-\theta)}-\frac{R}{2}\right)\right]-\frac{Rt_0}{2L}\sqrt{\frac{4L}{R^2C}-1},$$

and

$$K=\tfrac{1}{2}\varepsilon^{\frac{Rt_0}{2L}}\frac{\cos(\omega t_0-\theta)}{\cos(\beta t_0+\gamma)}$$

$$=\tfrac{1}{2}\varepsilon^{\frac{Rt_0}{2L}}\sqrt{\frac{C(2E\cos\omega t_0-2E_0+2L\omega I\sin(\omega t_0-\theta)-RI\cos(\omega t_0-\theta))^2}{(4L-R^2C)I}+\cos^2(\omega t_0-\theta)}.$$

Depending on which sign is taken for the square root, K may be either positive or negative. It is simpler to take the positive value, in which case $\cos(\beta t_0+\gamma)$ is positive also. If, however, the negative value of K is chosen, $\cos(\beta t_0+\gamma)$ is negative also, with a consequent change of π in the value of γ. The formula for γ has the corresponding ambiguity.

It is possible under certain circumstances for the current to be harmonic from the time of closing the circuit. In this case the starting term becomes zero. This requires that the circuit be closed at the instant when $\cos(\omega t-\theta)$ is zero, and that the initial potential difference between condenser terminals has the value

$$E_0=E\cos\omega t_0+L\omega I\sin(\omega t_0-\theta).$$

§ 47. Graphically the current may be represented by the resultant of four revolving vectors made up of two pairs. The first pair consists of two uniform circular vectors, revolving in opposite directions with angular velocity ω and with equal magnitudes. The second pair consists of two exponential spirals with angular velocities $\beta=\frac{R}{2L}\sqrt{\frac{4L}{R^2C}-1}$ in opposite directions and with equal magnitudes.

§ 48. If the resistance of the circuit is greater than or equal to $2\sqrt{\frac{L}{C}}$, the starting term loses its oscillatory character, and the formula for ψ becomes in the former case

$$\psi=K_1\varepsilon^{-\alpha_1 t}+K_2\varepsilon^{-\alpha_2 t},$$

where

$$\alpha_1=\frac{R}{2L}+\frac{R}{2L}\sqrt{1-\frac{4L}{R^2C}} \quad \text{and} \quad \alpha_2=\frac{R}{2L}-\frac{R}{2L}\sqrt{1-\frac{4L}{R^2C}},$$

and

$$\frac{d\psi}{dt}=-\alpha_1 K_1\varepsilon^{-\alpha_1 t}-\alpha_2 K_2\varepsilon^{-\alpha_2 t}.$$

At the time, t_0, of closing the circuit, the current is zero and

$$L\left[\frac{di}{dt}\right]_{t=t_0}=E\cos\omega t_0-E_0.$$

Substituting the values at the time $t=t_0$, we obtain

$$0=I[\cos(\omega t_0-\theta)-\psi_0]=I[\cos(\omega t_0-\theta)-K_1\varepsilon^{-\alpha_1 t_0}-K_2\varepsilon^{-\alpha_2 t_0}],$$

$$E\cos\omega t_0-E_0=LI[-\omega\sin(\omega t_0-\theta)+\alpha_1 K_1\varepsilon^{-\alpha t_0}+\alpha_2 K_2\varepsilon^{-\alpha_2 t_0}].$$

Substituting in this equation the value from the previous equation of $K_2\varepsilon^{-\alpha_2 t_0}$, $K_2\varepsilon^{-\alpha_2 t_0}=\cos(\omega t_0-\theta)-K_1\varepsilon^{-\alpha_1 t_0}$, we have

$$E\cos\omega t_0-E_0+L\omega I\sin(\omega t_0-\theta)$$
$$=LI[(\alpha_1 K_1-\alpha_2 K_1)\varepsilon^{-\alpha_1 t_0}+\alpha_2\cos(\omega t_0-\theta)],$$

and

$$K_1\varepsilon^{-\alpha_1 t_0}=\frac{E\cos\omega t_0-E_0+L\omega I\sin(\omega t_0-\theta)-\alpha_2 LI\cos(\omega t_0-\theta)}{(\alpha_1-\alpha_2)LI},$$

$$K_2\varepsilon^{-\alpha_2 t_0}=-\frac{E\cos\omega t_0-E_0+L\omega I\sin(\omega t_0-\theta)-\alpha_1 LI\cos(\omega t_0-\theta)}{(\alpha_1-\alpha_2)LI},$$

and

$$\psi=\frac{E\cos\omega t_0-E_0+L\omega I\sin(\omega t_0-\theta)-\alpha_2 LI\cos(\omega t_0-\theta)}{(\alpha_1-\alpha_2)LI}\varepsilon^{-\alpha_1(t-t_0)}$$
$$-\frac{E\cos\omega t_0-E_0+L\omega I\sin(\omega t_0-\theta)-\alpha_1 LI\cos(\omega t_0-\theta)}{(\alpha_1-\alpha_2)LI}\varepsilon^{-\alpha_2(t-t_0)}.$$

If R^2C equals $4L$, the starting term takes a third form,

$$\psi=(K_1+K_2t)\varepsilon^{-\frac{Rt}{2L}},$$

where K_1 and K_2 are constants depending on the conditions at the time t_0 of closing the circuit. Making the same assumptions as before, we shall find that

$$K_1=\varepsilon^{\frac{Rt_0}{2L}}\left[\left(1-\frac{Rt_0}{2L}\right)\cos(\omega t_0-\theta)\right.$$
$$\left.+t_0\left(\frac{E\cos\omega t_0-E_0}{LI}-\omega\sin(\omega t_0-\theta)\right)\right],$$

and

$$K_2=\varepsilon^{\frac{Rt_0}{2L}}\left[\frac{R}{2L}\cos(\omega t_0-\theta)-\frac{E\cos\omega t_0-E_0}{LI}-\omega\sin(\omega t_0-\theta)\right].$$

If the current is to be harmonic from the time of closing the circuit, evidently we must have in the last two cases, as in the first, § 46,

$$\cos(\omega t_0-\theta)=0,$$

and

$$E_0=E\cos\omega t_0+L\omega I\sin(\omega t_0-\theta).$$

GENERAL REMARKS

§ 49. It is evident also that these results for the oscillatory discharge of a condenser and for the current with simple harmonic e.m.f. when the starting term is oscillatory, might be expressed as the real part of an exponential spiral for the first case, and as the real part of the sum of a uniform circular quantity and an exponential spiral for the second.

For the oscillatory discharge we shall have

$$i = \text{real part}\left[\frac{2E_0\sqrt{C}}{\sqrt{4L-R^2C}}\varepsilon^{-\frac{Rt}{2L}\pm j\left(\frac{Rt}{2L}\sqrt{\frac{4L}{R^2C}-1}-\frac{\pi}{2}\right)}\right],$$

$$e = \text{real part}\left[\frac{2E_0\sqrt{L}}{\sqrt{4L-R^2C}}\varepsilon^{-\frac{Rt}{2L}\pm j\left(\frac{Rt}{2L}\sqrt{\frac{4L}{R^2C}-1}-\frac{\pi}{2}+\delta\right)}\right],$$

and for the current when the e.m.f. is simple harmonic,

$$i = \text{real part}\left[I\left(\varepsilon^{\pm j(\omega t-\theta)}-2K\varepsilon^{-\frac{Rt}{2L}\pm j(\beta t+\gamma)}\right)\right]$$

In general the angular velocity β of the spiral will be different from the angular velocity ω of the uniform circular component.

If β is a whole multiple of ω, or even nearly so, oscillatory surging of the current of corresponding frequency will occur if the e.m.f. has a harmonic of that frequency.

CHAPTER VII

COMPOUND HARMONIC CURRENT, E.M.F. AND POWER

§ 50. Periodic currents (or electromotive forces) which do not follow simple harmonic laws may be represented by means of revolving vectors. The most evident way of representing such a current is by the projection of a line whose length equals the maximum value of the current, the sign not being considered. The angular velocity of rotation may be varied as required, at times if necessary being reduced to zero or even reversed in sense. Such a method while involving no real difficulty, when viewed from a purely mechanical standpoint, in fact involves considerable difficulty when an attempt is made to express it in mathematical symbols.

USE OF FOURIER SERIES

§ 51. A better method is to resolve the periodic current into a number of harmonic terms. The periodic current, in other words, may be expressed as a Fourier series of the form,

$$i = I_0 + \sqrt{2}I_1 \cos(\omega t - \delta_1) + \sqrt{2}I_2 \cos(2\omega t - \delta_2)$$

$$+ \sqrt{2}I_3 \cos(3\omega t - \delta_3) + \text{etc.} + \sqrt{2}I_n \cos(n\omega t - \delta_n) + \text{etc.}$$

If preferred, the series may take the form

$$i=I_0+A_1 \sin \omega t+A_2 \sin 2\omega t+A_3 \sin 3\omega t+\text{etc.} +A_n \sin n\omega t+\text{etc.},$$
$$+B_1 \cos \omega t+B_2 \cos 2\omega t+B_3 \cos 3\omega t+\text{etc.} +B_n \sin n\omega t+\text{etc.}$$

The relations among the constants of the two equations are evident. The latter form of the equation has for our purposes little to commend it, and we shall not use it. The former form expresses the current as a constant plus the sum of the projections of lines of length $\sqrt{2}I_1$, $\sqrt{2}I_2$, $\sqrt{2}I_3$, etc., $\sqrt{2}I_n$, etc., where I_1, I_2, I_3, etc., I_n, etc., are the effective values of the components of the current.

In general, currents with which we shall deal may be represented with sufficient approximation by a very limited number of terms; and in most cases only the terms of odd order are present in currents of commercial circuits. The Fourier series in such cases reduces to

$$i=\sqrt{2}[I_1 \cos (\omega t-\delta_1)+I_3 \cos (3\omega t-\delta_3) +I_5 \cos (5\omega t-\delta_5)+\text{etc.}].$$

The constant term I_0 is only present in case the average value of the current differs from zero. The terms of even order are present if successive half waves differ in anything but sign, proper allowance being made for the constant term if present.

It is evident that the resultant revolving vector, in cases of this kind, is represented by a broken line, each part of which revolves with its own proper angular velocity.

Similar methods may be used to express a periodic e.m.f. Using the first form we may write

$$e=E_0+\sqrt{2}[E_1 \cos (\omega t-\lambda_1)+E_2 \cos (2\omega t-\lambda_2) +E_3 \cos (3\omega t-\lambda_3)+\text{etc.}+E_n \cos (n\omega t-\lambda_n)+\text{etc.}].$$

It is assumed that current and e.m.f. have equal frequencies.

§ 52. The power developed in the circuit is found by taking the product of e and i. We have in the product a number of terms of the form,

$$2E_qI_r \cos (r\omega t-\delta_r) \cos (q\omega t-\lambda_q).$$

If we substitute for these cosine products their equal,

$$E_qI_r [\cos ((r+q)\omega t-\delta_r-\lambda_q)+\cos ((r-q)\omega t-\delta_r+\lambda_q)],$$

and remember that the sum of a number of simple harmonic quantities of the same frequency is another simple harmonic quantity of that same frequency, the expression for p reduces to

$$\begin{aligned} p=E_0I_0+\sum_1^{\infty}E_nI_n \cos (\delta_n-\lambda_n)+P_1 \cos (\omega t-\beta_1) \\ +P_2 \cos (2\omega t-\beta_2)+P_3 \cos (3\omega t-\beta_3) \\ +\text{etc.}+P_n \cos (n\omega t-\beta_n)+\text{etc.} \end{aligned}$$

The average power is made up of the constant terms, for the variable terms (including all terms functions of ω) have an average value zero. The average power P is

$$\begin{aligned} P=E_0I_0+\sum_1^{\infty}E_nI_n \cos (\delta_n-\lambda_n)=E_0I_0+E_1I_1 \cos (\delta_1-\lambda_1) \\ +E_2I_2 \cos (\delta_2-\lambda_2)+\text{etc.}+E_nI_n \cos (\delta_n-\lambda_n)+\text{etc.} \end{aligned}$$

If, as is generally true of commercial circuits, the e.m.f. and current are approximately simple harmonic and if the only harmonics present are of odd order, the expression for average power will reduce to three or four terms whose sum is constant. The complete expression, practically considered, will reduce for the instantaneous power p to

$$\begin{aligned} p=P+P_2 \cos (2\omega t-\beta_2)+P_4 \cos (4\omega t-\beta_4) \\ +P_6 \cos (6\omega t-\beta_6)+\text{etc.} \end{aligned}$$

This may be expressed by an eccentric revolving vector whose origin is distant from the center of rotation by an amount P

POWER FACTOR

§ 53. For the power factor of such a circuit to be unity, it is necessary that the current and e.m.f. curves be precisely similar, i.e., the current at every instant must be in the same fixed proportion to the e.m.f. Mathematically expressed, this means that

$$E_0 : I_0 :: E_1 : I_1 :: E_2 : I_2 :: E_3 : I_3 :: \text{etc.} :: E_n : I_n :: \text{etc.},$$

and that $\lambda_1 = \delta_1$, $\lambda_2 = \delta_2$, $\lambda_3 = \delta_3$, etc., $\lambda_n = \delta_n$, etc. Under all other circumstances the product of the effective values of current and e.m.f. (I and E) will exceed the average power. This may be shown as follows:

The effective value I of the current, square root of mean square of i, is

$$I = \sqrt{I_0^2 + I_1^2 + I_2^2 + I_3^2 + \text{etc.}} = \sqrt{\sum_0^\infty I_n^2};$$

for as before in the expression for average power, the cross products involving different frequencies add nothing to the final result.

In the same way, the effective value E of the e.m.f. is

$$E = \sqrt{E_0^2 + E_1^2 + E_2^2 + E_3^2 + \text{etc.}} = \sqrt{\sum_0^\infty E_n^2}.$$

Let us suppose the various components of the current to be in the proportion

$$I_0 : I_1 : I_2 : I_3 : \text{etc.} :: 1 : \alpha : \beta : \gamma : \text{etc.},$$

and those of the e.m.f.

$$E_0 : E_1 : E_2 : E_3 : \text{etc.} :: 1 : \alpha_1 : \beta_1 : \gamma_1 : \text{etc.},$$

and that all the individual power factors are unity, *i.e.*, $\lambda_1=\delta_1$, $\lambda_2=\delta_2$, $\lambda_3=\delta_3$, etc. We then have

$$\begin{aligned} I &= I_0\sqrt{1+\alpha^2+\beta^2+\gamma^2+\text{etc.}},\\ E &= E_0\sqrt{1+\alpha_1^2+\beta_1^2+\gamma_1^2+\text{etc.}},\\ P &= E_0 I_0(1+\alpha\alpha_1+\beta\beta_1+\gamma\gamma_1+\text{etc.}), \end{aligned}$$

from which it follows that

$$\begin{aligned} \frac{E^2I^2-P^2}{E_0^2I_0^2} = {} & [1+\alpha^2\alpha_1^2+\beta^2\beta_1^2+\gamma^2\gamma_1^2+\text{etc.}\\ & +\alpha^2+\alpha_1^2+\beta^2+\beta_1^2+\gamma^2+\gamma_1^2+\text{etc.}\\ & +\alpha^2\beta_1^2+\alpha_1^2\beta^2+\alpha^2\gamma_1^2+\alpha_1^2\gamma^2+\text{etc.}+\beta^2\gamma_1^2\\ & \quad +\beta_1^2\gamma^2+\text{etc.}]\\ & -[1+\alpha^2\alpha_1^2+\beta^2\beta_1^2+\gamma^2\gamma_1^2+\text{etc.}\\ & +2\alpha\alpha_1+2\beta\beta_1+2\gamma\gamma_1+\text{etc.}\\ & +2\alpha\alpha_1\beta\beta_1+2\alpha\alpha_1\gamma\gamma_1+\text{etc.}+2\beta\beta_1\gamma\gamma_1+\text{etc.}]. \end{aligned}$$

Canceling like plus and minus quantities and combining the rest, we have

$$\begin{aligned} \frac{E^2I^2-P^2}{E_0^2I_0^2} = {} & (\alpha-\alpha_1)^2+(\beta-\beta_1)^2+(\gamma-\gamma_1)^2+\text{etc.}+(\alpha\beta_1-\alpha_1\beta)^2\\ & +(\alpha\gamma_1-\alpha_1\gamma)^2+\text{etc.}+(\beta\gamma_1-\beta_1\gamma)^2+\text{etc.} \end{aligned}$$

As the right side of the equation is the sum of squares, it cannot be negative, and can be zero only if $\alpha=\alpha_1$, $\beta=\beta_1$, $\gamma=\gamma_1$, etc. Under all other circumstances we must have EI greater than P. If the individual power factors are less than unity, the value of P will be smaller still. We therefore see that for the power factor of the effective current to be unity, the current and e.m.f. must have precisely similar form.

§ 54. It is well known that hysteresis modifies the form of the current curve, so that it cannot be as a rule of the same form as the e.m.f. curve. It is for this reason usually impossible precisely to obtain unity power factor in the case of a synchronous motor whose field has been adjusted for maximum power factor; for no one field excitation can bring all instantaneous values of current and e.m.f. into a constant ratio.

CHAPTER VIII

INTERLINKED CIRCUITS, MUTUAL INDUCTION

§ 55. When two circuits are linked together by means of the magnetic lines of force, due to currents in both or either circuit, we may observe the phenomenon of electromotive forces due to mutual inductance. This phenomenon was first observed by Joseph Henry. The phenomenon of self inductance was discovered by Michael Faraday, who gave to the world the invaluable conception of lines of force to explain inductive phenomena, whether electric or magnetic in origin. Faraday was not a mathematical physicist and his famous researches with respect to lines of force were not appreciated until put into mathematical language by Maxwell. Unfortunately for our present electromagnetic terminology, Maxwell saw fit to use the expression *lines of induction* in place of Faraday's lines of force which we commonly represent by Φ (for the total flux) and **B** (for flux per unit area). Maxwell also unfortunately gave the expression *lines of force* a new meaning, denoting by it field strength, for which we use the symbol **H** (for unit area). Many writers by error use **H** to represent Faraday's lines of force in electromotive force formulæ. It is evident that change of flux (**B**), not field strength (**H**), determines the inductive electromotive force. Such writers use Φ to represent the surface integrals of **H** and **B** indifferently. The writer in company with many others favors holding to Faraday's expression lines of force to represent flux (not field strength).

OHM'S LAW EXTENDED TO MUTUALLY INDUCTIVE CIRCUITS

§ 56. When two electric circuits are interlinked by lines of force, we have a new term in the law of Ohm (extended to variable conditions). The expression for the electromotive force e_1 impressed on the primary circuit, designated by the subscript 1, becomes

$$e_1 = R_1 i_1 + L_1 \frac{di_1}{dt} + M \frac{di_2}{dt}.$$

In this expression R_1 and L_1 are the resistance and self inductance of the primary coil, i_1 is the current in the primary circuit, M is the mutual inductance between the two circuits and i_2 is the current in the secondary circuit. The e.m.f. produced in the secondary circuit, because of rate of change in the primary circuit, is equal to $-M\frac{di_1}{dt}$. If the secondary circuit is closed and a current i_2 is produced, a part of this e.m.f. is lost in ohmic drop of potential $i_2 R_2$ and electromotive force of self induction $L_2 \frac{di_2}{dt}$, leaving available at the terminals the remainder e_2, impressed by the secondary on its external circuit. We have then

$$e_2 = -M\frac{di_1}{dt} - R_2 i_2 - L_2 \frac{di_2}{dt}.$$

The difference in the form of the two equations is due to the conventional agreement to consider the primary e.m.f. to be *applied to* the primary, and the second e.m.f. to be *applied by* the secondary.

§ 57. The primary and secondary e.m.f.'s may be expressed in terms of *Faraday's* lines of force also. Calling the flux linking with the primary ϕ_1, that linking with the secondary ϕ_2, and that linking with both, ϕ, with maximum values of Φ_1, Φ_2, and Φ respectively, and designating by N_1 and N_2 the turns of wire in the two coils, we shall have

$$e_1 = R_1 i_1 + N_1 \frac{d\phi_1}{dt},$$

$$e_2 = -N_2 \frac{d\phi_2}{dt} - R_2 i_2.$$

If the primary and secondary circuits were so closely related that $\phi_1 = \phi_2 = \phi$, all lines of flux must link with every turn of both coils. While in fact this is an impossible assumption, we do find it closely approached in good commercial transformers with moderate loads.

§ 58. The method of revolving vectors may be used to represent the electromotive forces, currents and fluxes. The application is not difficult if they follow harmonic laws. If all are simple harmonic, we are dealing with an ideal transformer. In the case of practical transformers, we find that even if the electromotive forces are simple harmonic, the currents and fluxes will follow more complicated laws.

FARADAY'S RING

§ 59. Let us consider the simplest form of transformer, a Faraday ring. Suppose the core built up of thin sheets of extremely soft iron of high permeability μ. We shall assume this permeability to be constant for all values of ϕ. We shall also assume that no flux leaves the ring, and that the primary and secondary windings have extremely low resistance, and are so close to the ring and so well-intermingled that no flux exists outside the ring. We shall then have an ideal transformer. Let the core have a permeability μ, an average length K and a cross-section

A. As before there are N_1 primary and N_2 secondary turns, and the primary and secondary currents are i_1 and i_2. Let us first suppose i_2 is zero, the condition of an open secondary circuit. The magnetic field has been supposed to satisfy the solenoidal condition, that all lines of field intensity keep within the core, like water flowing in a pipe.[1] The magnetic reluctance of the core is $\frac{K}{A\mu}$. The magnetomotive force is

$$M.M.F. = 4\pi N_1 i_1,$$

and it produces in the core a flux of lines of force (Faraday's)

$$\phi = \frac{4\pi N_1 A \mu i_1}{K},$$

with a flux intensity per unit cross-section of the core,

$$\mathbf{B} = \frac{4\pi N_1 \mu i_1}{K}.$$

It has been assumed that the current is in C.G.S. units. If i_1 is expressed in amperes, we must write $10K$ in the place of K in both equations.

The inductive electromotive force due to rate of change of flux is as before $-N_1\frac{d\phi}{dt}$ in the primary, and $-N_2\frac{d\phi}{dt}$ in the secondary. We therefore have

$$e_1 = R_1 i_1 + N_1\frac{d\phi}{dt} = R_1 i_1 + \frac{4\pi N_1^2 A\mu}{K}\frac{di_1}{dt} \text{ (open secondary)},$$

and

$$e_2 = -N_2\frac{d\phi}{dt} = -\frac{4\pi N_1 N_2 A\mu}{K}\frac{di_1}{dt} \text{ (open secondary)}.$$

[1] The words *solenoid* and *solenoidal* are derived from the Greek word for pipe or channel, and are intended to convey the idea of a flux keeping within its channel.

If the secondary circuit is closed, ϕ will depend on both i_1 and i_2, and we shall have

$$e_1 = R_1 i_1 + \frac{4\pi N_1 A \mu}{K}\left(N_1 \frac{di_1}{dt} + N_2 \frac{di_2}{dt}\right),$$

and

$$e_2 = -R_2 i_2 - \frac{4\pi N_2 A \mu}{K}\left(N_1 \frac{di_1}{dt} + N_2 \frac{di_2}{dt}\right).$$

The conclusions reached are valid only when μ is constant and the flux solenoidal, i.e., there is no leakage of flux.

CONCERNING LINES OF FORCE

§ 60. To prove that it is flux (Faraday's lines of force), not field strength (Maxwell's lines of force), which determines electromotive force, we may consider two transformers built with precisely similar dimensions, their only difference being that one has a well-laminated soft iron core having a permeability say 3000, the other having a wooden core of permeability 1. Let the secondaries be connected to high-resistance voltmeters, but otherwise have no load. The secondary currents are negligible. Connect the primaries of the two transformers in series and apply an e.m.f., which will cause full secondary voltage in the former. The secondary voltage of the latter will be $\frac{1}{3000}$ of that of the former. The currents in the two primaries are alike and they produce equal field strength **H** in both cores. If it were rate of field strength change which determines e.m.f., the transformers would have equal secondary voltages. We find on the contrary that the secondary voltages are in the ratio of the number of Faraday's lines of force. From this we conclude that it is **B**, not **H**, which determines electromotive force.

RATIO OF TRANSFORMATION

§ 61. We showed in § 57 that in a transformer

$$e_1 = R_1 i_1 + N_1 \frac{d\phi_1}{dt},$$

and

$$e_2 = -N_2 \frac{d\phi_2}{dt} - R_2 i_2.$$

For small currents, $R_1 i_1$ and $R_2 i_2$ will be small. If they may be neglected, and if there is negligible loss of flux, so that we may consider ϕ_1 and ϕ_2 equal, we obtain the approximate ratio

$$e_1 : e_2 :: N_1 : -N_2.$$

If e_1 follows a simple harmonic law, we have

$$\phi = \Phi \cos (\omega t - \frac{\pi}{2}) = \Phi \sin \omega t,$$

and neglecting the ohmic drop $R_1 i_1$ and $R_2 i_2$, we shall have

$$e_1 = \sqrt{2} E_1 \cos \omega t = N_1 \omega \Phi \cos \omega t,$$

and

$$e_2 = \sqrt{2} E_2 \cos (\omega t - \pi) = -N_2 \omega \Phi \cos \omega t;$$

and finally for the ideal ratio of transformation of electromotive forces (effective values), we shall have

$$E_1 : E_2 :: N_1 : N_2.$$

In practical transformers there is always some loss of flux due to magnetic leakage, and $R_1 i_1$ and $R_2 i_2$ cannot be neglected. For these reasons the secondary electromotive force generally falls below its ideal value.

§ 62. In our ideal transformer we have assumed the core well laminated and of very high permeability. If the secondary circuit is open the primary will require an extremely small current to produce the necessary flux. Had the core not been laminated, eddy currents generated in the core would have required more primary current. Also had the permeability of the core been low, a greater magnetizing current would have been required.

If the secondary is delivering a normal current, the magnetizing effect of the primary and secondary currents will practically offset one another; for it has been assumed that owing to the high permeability of the core, very little magnetizing current is required. The magnetizing effect is proportional to the aggregate ampere-turns. We therefore have i_1N_1 practically equal to $-i_2N_2$ at all times, or for effective values, neglecting magnetizing current,

$$I_1 : I_2 :: N_2 : N_1,$$

TRANSFORMER DIAGRAMS, LAGGING CURRENT

§ 63. The phase relation of current and e.m.f., primary and secondary, depends on the exterior portion of the secondary circuit. If this portion of the circuit is a non-inductive resistance, the secondary current and e.m.f. will be in the same phase. The same is true also for any arrangement giving unit power factor. If the secondary current is out of phase with its e.m.f. we may have either lead or lag.

In any case a revolving vector diagram may be used to represent the facts. To illustrate a case in which the current lags behind the e.m.f., let us draw a line OC_2 to represent the secondary ampere-turns I_2N_2 and a line OB_2 to represent the secondary volts-per-turn $\frac{E_2}{N_2}$ in their proper phase relation (Fig. 20). Draw parallel to OC_2 a line B_2A_2 to represent ohmic drop per secondary turn $\frac{R_2I_2}{N_2}$. Then OA_2 will represent the e.m.f. per secondary term produced by the varying flux. At right angles to OA_2 (90° in advance) draw OF to represent the flux Φ. Opposite to OA_2, and equal to it, draw OA_1 to represent the part of the applied e.m.f. required per turn to balance the induced e.m.f. Draw the line OC_1 equal and opposite to OC_2 to represent the primary ampere-turns I_1N_1, which were assumed to equal the secondary ampere-turns. Draw the line A_1B_1 parallel to OC_1 to represent the primary ohmic drop per turn $\frac{R_1I_1}{N_1}$, and last draw the line OB_1, which represents the volts-per-turn which must be applied to the primary by some external source.

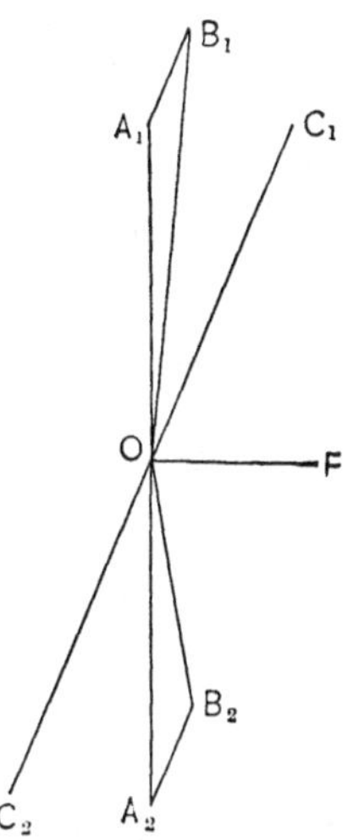

Fig. 20.

Similar quantities must be drawn to the same scale, but the ampere-turns need not be drawn to the same scale as the volts-per-turn. The reason for representing ampere-turns rather than amperes, and volts-per-turn instead of volts, is to keep the lines of reasonable length. Otherwise in a transformer with high ratio of transformation, similar quantities would be represented by lines of very inconvenient lengths.

EXCITING CURRENT, CORE LOSSES

§ 64. If the core losses due to hysteresis and eddy currents are not negligible, the primary current must be increased to provide for these losses. This additional component is commonly called the magnetizing or exciting current. Strictly speaking, the magnetizing current is only the part that provides for hysteresis. The magnetizing current is never simple harmonic, as hysteresis always produces a certain distortion in the wave form. We may as a rule ignore the higher harmonics, as they are practically negligible in comparison with the load currents. This may be illustrated as follows: If I_1 represents the whole primary current, with components A_1 of fundamental frequency and A_3 of three times as great frequency, and if higher terms are negligible, we have

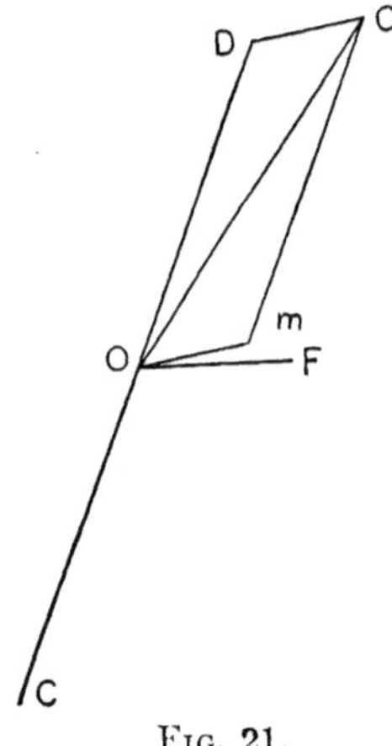

FIG. 21.

$$I_1 = \sqrt{A_1^2 + A_3^2} = A_1\left(1 + \tfrac{1}{2}\frac{A_3^2}{A_1^2} - \tfrac{1}{8}\frac{A_3^4}{A_1^4} + \text{etc.}\right),$$

from which it will be seen, for example, that if $A_3 = 0.1A_1$, we shall have $I_1 = 1.005\ A_1$.

To provide power for hysteresis and eddy currents, the exciting current must be ahead of the flux in phase. In Fig. 21 we represent the ampere-turns of the exciting current by the line Om. The line OD is equal and opposite to OC_2; and OC_1, the resultant of OD and Om, represents the total primary ampere-turns. Put in another way, we may say that Om is the resultant of OC_1 and OC_2.

EFFECT OF THE FLUX LEAKAGE

§ 65. In case some of the lines of force linking with the primary coil do not link with the secondary coil, but pass outside the core, we have a condition of affairs practically equivalent to considering the useful flux as linking with both coils, and the leakage flux linking with a choke coil in series in the primary circuit and having a number of turns equal to those of the primary. The choke coil is supposed to have no resistance. The potential drop in the choke coil will lead the ohmic drop in the primary by 90°. The flux, which links with a portion only of the turns in either coil, is equivalent to a smaller amount linking with all and having the same total amount of flux-turns.

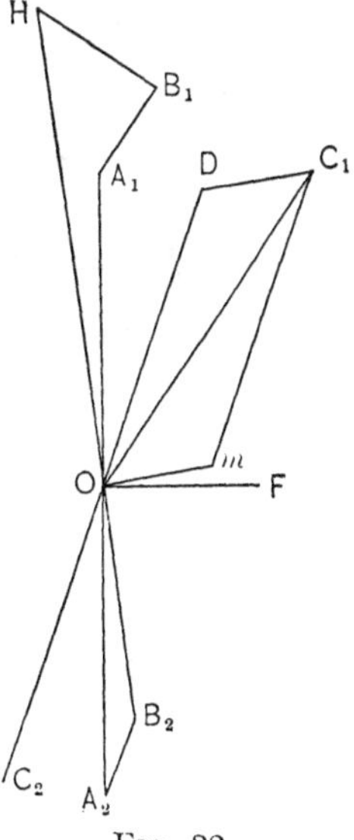

FIG. 22.

In Fig. 22, B_1H, which leads OC_1 (primary ampere-turns) by 90°, represents the primary volts-per-turn consumed by flux leakage. A_1B_1 is the ohmic drop per primary turn. OH is the total applied e.m.f. per primary turn.

TRANSFORMER EQUATIONS

§ 66. The relations among the quantities represented by Fig. 22 may be expressed in the form of equations. Let us assume that the external portion of the secondary circuit has an impedance r_2+jx_2; and let us assume that the flux linking with the primary circuit, but not with the secondary, is ϕ', with maximum value Φ'. The exciting current is m. The other symbols have the same significance as before. We then have for the revolving vectors,

$$\dot{I}_2(r_2+jx_2)=\dot{E}_2,$$

$$\omega N_2\dot{\Phi}_2=j\sqrt{2}(\dot{E}_2+\dot{I}_2R_2),$$

$$\dot{I}_1=\dot{m}-\frac{N_2}{N_1}\dot{I}_2,$$

$$\sqrt{2}\dot{E}_1=j\omega N_1(\dot{\Phi}_2+\dot{\Phi}')+\sqrt{2}\dot{I}_1R_1.$$

TRANSFORMER DIAGRAMS, LEADING CURRENT

§ 67. If the secondary current leads its electromotive force, the proper phase relations are to be taken into consideration in making the diagram.

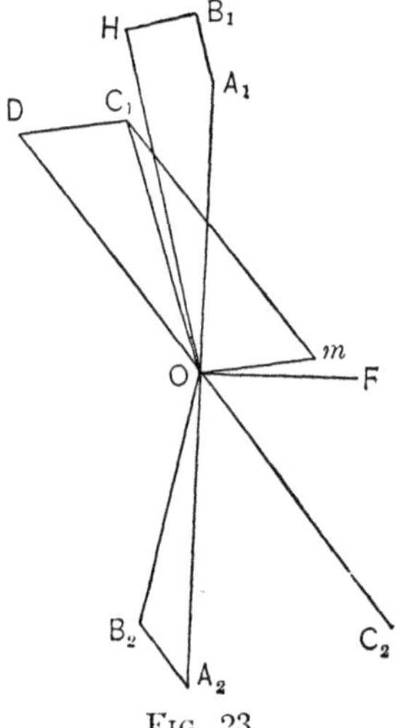

Fig. 23,

Fig. 23 illustrates the case of a leading secondary current. OC_2 represents the secondary ampere-turns, OB_2 the terminal secondary volts-per-turn, OC_1 the primary ampere-turns, and OH the primary terminal volts-per-turn. The ohmic drop, volts-per-turn, is represented by A_2B_2, in the secondary, and by A_1B_1, in the primary. The inductive drop in the primary, volts-per-turn, is represented by B_1H. Because of the phase relations, the inductive drop in Fig. 23 is not a drop at all, but rather a rise; for the terminal volts-per-turn OH are actually less than the amount OB_1, which would have been required for the same conditions of the secondary with no magnetic leakage.

DIFFICULTY FOUND IN EXPONENTIAL EXPRESSION

§ 68. To express the magnitudes represented in the previous equations and diagrams in terms of exponentials, the phase relations must all be determined. If, for example, it is desired to express the secondary current in terms of the primary electromotive force and the constants of the circuit, we shall find the exponential expression to be very complicated. On the whole it is better in a numerical problem to proceed through the series of equations given in § 66, using numerical values. Practically it is difficult to determine the leakage flux and the magnetizing current. The reason is because the permeability of the core is not constant, and the leakage flux, therefore, is higher at larger loads. The difficulties are, however, not insuperable.

CONCLUSION

§ 69. It is beyond the scope of this small book to consider all the alternating current problems to which the method of revolving vectors may be applied. If the reader has become well enough acquainted with the method to feel confidence in applying it when occasion arises, the author's purpose in writing on the subject has been accomplished.

INDEX

www.ingramcontent.com/pod-product-compliance
Lightning Source LLC
LaVergne TN
LVHW010106110826
845155LV00028B/512

* 9 7 8 1 9 3 4 9 3 9 2 9 1 *